Planet Earth
GLACIER

Other Publications:

AMERICAN COUNTRY
VOYAGE THROUGH THE UNIVERSE
THE THIRD REICH
THE TIME-LIFE GARDENER'S GUIDE
MYSTERIES OF THE UNKNOWN
TIME FRAME
FIX IT YOURSELF
FITNESS, HEALTH & NUTRITION
SUCCESSFUL PARENTING
HEALTHY HOME COOKING
UNDERSTANDING COMPUTERS
LIBRARY OF NATIONS
THE ENCHANTED WORLD
THE KODAK LIBRARY OF CREATIVE PHOTOGRAPHY
GREAT MEALS IN MINUTES
THE CIVIL WAR
COLLECTOR'S LIBRARY OF THE CIVIL WAR
THE EPIC OF FLIGHT
THE GOOD COOK
WORLD WAR II
HOME REPAIR AND IMPROVEMENT
THE OLD WEST

For information on and a full description of any of
the Time-Life Books series listed above, please call
1-800-621-7026 or write:
    Reader Information
    Time-Life Customer Service
    P.O. Box C-32068
    Richmond, Virginia 23261-2068

This volume is one of a series that examines
the workings of the planet earth, from the
geological wonders of its continents to the
marvels of its atmosphere and its ocean depths.

Cover
Berg Glacier forms a magnificent frozen cascade
down the rugged 12,000-foot flanks of Mount
Robson in British Columbia. One of thousands
of glaciers in the Canadian Rockies, it
periodically calves icebergs into the lake below.

Planet Earth

GLACIER

By Ronald H. Bailey
and The Editors of Time-Life Books

Time-Life Books, Alexandria, Virginia

Time-Life Books Inc.
is a wholly owned subsidiary of

TIME INCORPORATED

FOUNDER: Henry R. Luce 1898-1967

Editor-in-Chief: Henry Anatole Grunwald
Chairman and Chief Executive Officer: J. Richard Munro
President and Chief Operating Officer: N. J. Nicholas Jr.
Chairman of the Executive Committee: Ralph P. Davidson
Corporate Editor: Ray Cave
Executive Vice President, Books: Kelso F. Sutton
Vice President, Books: George Artandi

TIME-LIFE BOOKS INC.

EDITOR: George Constable
Director of Design: Louis Klein
Director of Editorial Resources: Phyllis K. Wise
Acting Text Director: Ellen Phillips
Editorial Board: Russell B. Adams Jr., Dale M.
Brown, Roberta Conlan, Thomas H. Flaherty, Donia
Ann Steele, Rosalind Stubenberg, Kit van Tulleken,
Henry Woodhead
Director of Photography and Research:
John Conrad Weiser

PRESIDENT: Christopher T. Linen
Executive Vice President: John M. Fahey Jr.
Senior Vice Presidents: James L. Mercer,
Leopoldo Toralballa
Vice Presidents: Stephen L. Bair, Ralph J. Cuomo,
Terence J. Furlong, Neal Goff, Stephen L. Goldstein,
Juanita T. James, Hallett Johnson III, Robert H.
Smith, Paul R. Stewart
Director of Production Services: Robert J. Passantino

PLANET EARTH

EDITOR: Thomas A. Lewis
Senior Editor: Anne Horan
Designer: Donald Komai
Chief Researcher: Pat S. Good

Editorial Staff for *Glacier*
Picture Editor: Carol Forsyth Mickey
Writers: Lynn Addison, William C. Banks,
Adrienne George, Kathy Kiely, John Newton
Researchers: Elizabeth B. Friedberg (principal),
Kristin Baker, Megan H. Barnett, Susan S. Blair,
Therese A. Daubner, Patricia A. Kim,
Oliver G.A.M. Payne
Assistant Designer: Susan K. White
Copy Coordinators: Allan Fallow, Victoria Lee,
Bobbie C. Paradise
Picture Coordinator: Donna Quaresima
Editorial Assistant: Annette T. Wilkerson

Editorial Operations
Copy Chief: Diane Ullius
Editorial Operations: Caroline A. Boubin (manager)
Production: Celia Beattie
Quality Control: James J. Cox (director)
Library: Louise D. Forstall

Correspondents: Elisabeth Kraemer-Singh (Bonn);
Maria Vincenza Aloisi (Paris); Ann Natanson
(Rome). Valuable assistance was also provided by:
Helga Kohl, Angelika Lemmer (Bonn); Caroline
Alcock, Lesley Coleman (London); Carolyn Chubet,
Donna Lucey (New York); Dag Christensen
(Oslo); John Scott (Ottawa); M. T. Hirschkoff
(Paris); Bogi Agustsson (Reykjavik); Mimi Murphy,
Ann Wise (Rome); Janet Zich (San Francisco);
Traudl Lessing (Vienna).

TIME-LIFE is a trademark of Time Incorporated U.S.A.

Library of Congress Cataloguing in Publication Data
Bailey, Ronald H.
  Glacier.
  (Planet earth; 5)
  Bibliography: p.
  Includes index.
  1. Glaciers. I. Time-Life Books. II. Title.
III. Series.
GB2403.2.B34   551.3'12   82-781
ISBN 0-8094-4316-3   AACR2
ISBN 0-8094-4317-1 (lib. bdg.)
ISBN 0-8094-4318-X (mail order ed.)

THE AUTHOR

Ronald H. Bailey is a freelance author and journalist who was formerly a senior editor of *Life.* He has written many previous volumes for Time-Life Books, dealing with subjects that range from World War II to the functioning of the human brain.

THE CONSULTANTS

Colin Bull, Professor of Geology and Mineralogy at Ohio State University, was formerly director of its Institute of Polar Studies and is now Dean of the College of Mathematical and Physical Sciences. He has studied glaciers in the Arctic and organized or led numerous expeditions to the Antarctic Peninsula and to the ice-free areas of Victoria Land, where Bull Pass and Bull Lake are named for him.

R. Quincy Robe is with the United States Coast Guard Research and Development Center in Groton, Connecticut. He has served as the International Ice Patrol scientist on a number of West Greenland glacier surveys and has participated in oceanographic research cruises in the Grand Banks and Baffin Bay.

CONTENTS

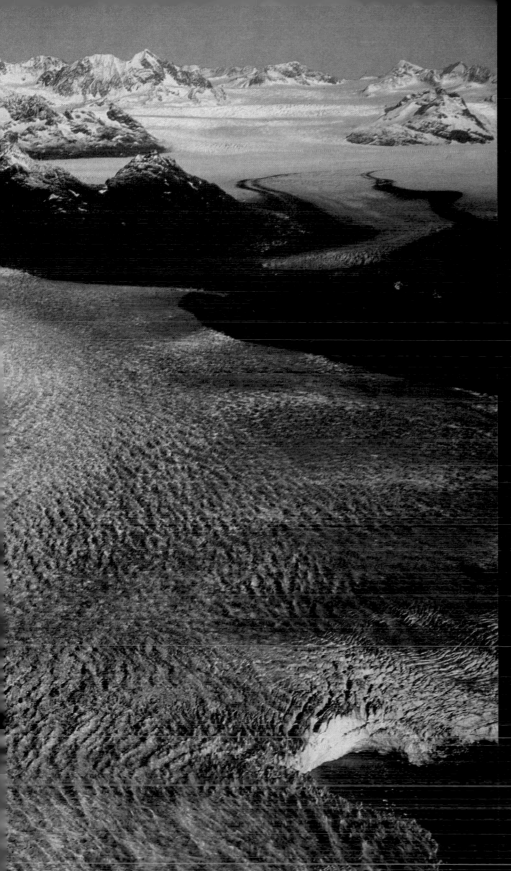

The mightiest rivers on earth are frozen solid and normally flow at slower than a snail's pace. Yet these great icy currents—called glaciers—are among the most spectacular and powerful agents of nature. Often thousands of feet wide and many miles long, they entomb millions of acres of land as they creep relentlessly through some of the world's wilderness areas.

Mountain glaciers, such as the ones shown here and on the following pages, display a dazzling variety of forms and colors—the result of their unceasing battle with weather and rugged terrain. But it is an uneven contest; however the elements may shape a glacier, the ice routinely prevails to shatter and gouge even the sturdiest rock formations. The resulting debris is pushed and dragged beneath the glacier and along its surface, eventually to be deposited in a titanic reshaping of the landscape.

Though present-day glaciers are mere vestiges of their former expanses, they cover almost 11 per cent of the world's land. Ice sheets as much as 14,000 feet thick smother all of Antarctica and most of Greenland, the world's largest island. The weight of these two immense ice sheets—estimated at 30 trillion tons—literally dents the earth; if the Greenland ice were to disappear, the island would eventually rebound some 2,000 feet. Moreover, glacial ice contains close to 90 per cent of the world's fresh water aboveground, and the frigid air above the vast ice sheets profoundly influences worldwide weather patterns.

In modern times, the study of glaciers has become a science in its own right, and great progress has been made in understanding glacial mechanics and effects. But for many students of these ponderous torrents of ice, awe underlies analysis. As American author Mark Twain wrote after a tour of Swiss glaciers in the late 1870s, "A man who keeps company with glaciers comes to feel tolerably insignificant by and by."

Extending into a deep fjord, the Columbia Glacier spreads a textured carpet of ice over the southern Alaska coastline near the oil-shipping port of Valdez. The glacier's retreat threatens to launch thousands of dangerous icebergs into nearby tanker lanes.

During an assault on Mount Everest in 1972, a British mountaineering team carefully skirts a menacing crevasse that splits the great Khumbu Glacier at an altitude of 20,000 feet on the peak's western flank. Crevasses, sometimes as deep as 100 feet, are caused by stress in the ice, and since these gaping cracks are often hidden beneath a delicate cover of snow, travel across a glacier is extremely dangerous.

Exploring an ice cave within the Muir Glacier in southeastern Alaska, a researcher examines an outcrop of ice so dense that it has turned a brilliant blue. Carved by meltwater and then sculpted into a scalloped pattern by seasonal temperature fluctuations, glacier caves are as treacherous as they are magnificent; during warm periods, walls may collapse without warning, and flash flooding may occur.

Meltwater from the 100-square-mile Matanuska
Glacier in south-central Alaska scours
out a dramatic overhang at the terminus of
the ice. The rocks below were deposited
when the glacial ice melted; atop the overhang
more debris is about to tumble to earth.
Horizontal layers of compacted snow and ice
are visible along the glacier's curved walls;
the vertical lines imparting a tile pattern
resulted when water froze in tiny crevasses.

Descending majestically from the slopes
of the Andes, the Moreno Glacier in Argentina
blankets a glacier-carved lake at its foot
with a miniature ice pack of broken fragments.

"GOD'S GREAT PLOUGH"

The glacier had no name, only a number. Perched on the steep northwest face of Peru's highest peak, the 22,205-foot Mount Huascarán, it had been there for as long as people could remember, advancing and retreating with the seasons—creeping ahead a few inches a day when nourished by the winter snows and shrinking back in the heat of summer. It was one of hundreds of glaciers that dappled the Andes, and Peruvian geologists had only recently given official recognition to this particular mass of ice, labeling it Glacier No. 511.

People in the Huaylas valley below Mount Huascarán tended their flocks of sheep on the lower slopes or raised fruit, grain and vegetables in the fertile bottom lands along the Santa River. Occasionally, an intrepid climber would scale Mount Huascarán to chip off a block of ice, then wrap it in grass to retard melting and carry it down on his back to sell to the restaurants of Ranrahirca and the other small towns of the valley. Otherwise, the local inhabitants paid little heed to the glacier. And on January 10, 1962, when Ricardo Olivera switched on the diesel generator at precisely 6 p.m. to provide Ranrahirca's nightly five hours of electricity, the looming white mass was merely a distant glimmer in the setting sun.

As dusk gathered about the summit of the mountain, however, Glacier 511 was beginning to stir. Its weather-cracked tongue was pitched near the edge of a rocky cliff. The glacier had occupied this position for years, and though small blocks of ice had frequently cracked away to go tumbling down the cliff, the glacier itself had remained relatively stable. But now a great mass of ice suddenly lurched into motion. Exactly what happened to awaken the somnolent glacier no one knows. Surface meltwater had almost certainly seeped through to the bottom of the ice mass, lubricating its footing over a vast area. In addition, some geologists believe, rockslides may have cascaded onto the glacier from the peak above, causing it to surge sharply forward.

At 6:13 p.m., 13 minutes after Ricardo Olivera had turned on the lights of Ranrahirca, Glacier 511 shuddered, and an enormous mass of ice broke loose. The ice—200 yards long and nearly half a mile wide—hurtled down the cliff and smashed into the gorge 3,000 feet below the glacier. The impact of millions of tons of ice raised a powdery white cloud that gleamed through the twilight; then came the noise, echoing across the Huaylas valley with a roar, someone later recalled, "like that of 10,000 beasts."

Nine miles away, in Ranrahirca, many of the town's 2,500 residents were just sitting down for the evening meal. Lamberto Guzmán Tapia had ar-

Several glaciers flow from two massive icecaps into the upper reaches of western Greenland's "Eternal Fjord." The 50-mile-long fjord, with walls as high as 5,500 feet, was carved out by glacial ice over a period of two million years.

Weakened by deep crevasses and warm-weather melting, a wall of ice topples from the glacial icecap atop 22,205-foot Mount Huascarán—Peru's highest mountain—at 6:13 p.m. on January 10, 1962. Wrenching huge rocks from the cliff face as it fell, the mass of some three million tons thundered onto the lower tongue of the glacier 3,000 feet below, then cascaded down the glacier's steep 35-degree slope toward the populous Huaylas valley.

After accelerating across the surface of the two-mile-long glacier, the avalanche roars through a ravine at more than 65 mph, forming a wall of ice, rock and mud 150 feet high. By this time, friction generated by the plunging mass had melted thousands of tons of pulverized ice, giving the avalanche a viscous appearance.

Only two minutes old but already bloated with debris scoured from the walls and floor of the twisting ravine, the avalanche obliterates the first of nine villages in its path. Though the original mass had more than doubled, the decrease in the slope and the repeated collisions with the ravine walls slowed the torrent to 40 mph as it entered the populated areas.

Ten miles from its starting point, the ice flood spreads out over the alluvial plain and begins to engulf the town of Ranrahirca, crushing it like a gigantic steam roller. When the avalanche came to rest two minutes later, more than 2,400 townspeople were dead.

rived belatedly at his aunt's house where 40 guests were clapping their hands and singing *huaynos,* old Peruvian folk songs. Tapia was an avid mountain climber, and when he heard the distant roar above the party's happy din, he knew instantly what it was. "Avalanche!" he shouted.

The avalanche was just beginning. When the ice tumbled into the gorge, it already contained rock debris gathered by Glacier 511 from its own bed. Now, as the ice bounced against the walls of the gorge in a crazy zigzag (geologists later found five separate points of impact), it plowed up granite chunks as big as houses and carried them along. Down the gorge raced this runaway mass of rock and ice, gathering momentum and more debris. At speeds exceeding 65 miles per hour it climbed foothills as high as 275 feet, leaving a 6,000-ton boulder perched atop one crest, and stirred hurricane-like winds on either flank. It swept up everything in its path: topsoil, trees, sheep and llamas grazing on the mountainside and, all too soon, the human beings who had paid little attention to the glacier without a name.

According to geologists who later reconstructed the timetable of terror launched by Glacier 511, the ice avalanche reached full force in barely two minutes. At 6:15 it took the first human lives, crushing the little Quechua Indian village of Yanamachico and several nearby mountain hamlets, leaving nearly 1,000 dead. Then, at 6:16, the avalanche reached the more populated bottom lands of the valley. The slope was more gradual there, and the avalanche slowed somewhat, following a river valley at about 40 miles per hour. Only a high bank alongside the avalanche's right flank kept the deadly mass from reaching the city of Yungtay, a short distance west.

But the glacial avalanche was bearing down upon Ranrahirca. What had started as some 2.5 million cubic yards of ice was now a mass of ice, rock and mud five times greater. The mass was a mile wide and more than 40 feet thick. Just above Ranrahirca, a 60-year-old widow named Zoila Chrishna Angel saw it "rushing at us like the end of the world." Voices shouted, "Run! Run!" "I could not run," she said later. "I could not move. I could not speak."

The avalanche roared by, leaving her unharmed as she stood trembling on the hillside. And at 6:18, the ice and debris plowed through Ranrahirca at 40 miles an hour. Red-tiled roofs, cobblestone streets, 2,400 residents—all were entombed under as much as 60 feet of glacial rubble. Two minutes later, about a mile beyond Ranrahirca, the terrible brown-and-white mass bridged the Santa River, climbed 100 feet up the opposite bank and, its fury finally spent, came to rest.

The avalanche spawned by Glacier 511 had raced 10 miles in eight minutes. An estimated 4,000 people from nine villages were dead, most of them buried beneath the ice, mud and rock, though bodies were found as far away as 100 miles downstream, where the Santa River emptied into the Pacific Ocean.

Among the 100 or so residents of Ranrahirca who survived was mountaineer Lamberto Tapia, who had fled uphill out of the path of the avalanche. The electrician Olivera was spared because the power plant was on the outskirts of town, just beyond the swath of destruction. Shortly after the avalanche had roared through, killing Olivera's wife and children—he lost 27 relatives in all—he sadly turned off the generator. But then Olivera had second thoughts and flipped the switch back on. In a few houses nearby—all that remained of Ranrahirca—the lights went on again.

Fortunately, ice avalanches triggered by glaciers are rare; in terms of lives lost, the event on Peru's Mount Huascarán was the worst ever recorded. But the cataclysm that brought such sorrow to Ranrahirca demonstrated—trag-

One week after the avalanche, survivors pick their way along a rickety wooden catwalk across the mile-wide river of debris. Some nine million cubic yards of ice, rock and mud covered more than 860 acres of the valley floor to depths of as much as 60 feet. Stretching 10 miles from the glacier to the bottom of the valley, the deluge of debris created an icy sepulcher for as many as 4,000 people.

ically and to an extreme degree—the awesome power of ice to affect the earth and its inhabitants.

Ranging in size from insignificant tongues of ice, only a few hundred yards long in the Rocky Mountains of North America, to the continental sheet of ice more than five million miles square that blankets Antarctica, glaciers cover no less than 10 per cent of the earth's land surface. They can be found on every continent, even on the tropical island of New Guinea. But the places cold and snowy enough to sustain glaciers—high mountain ranges and the polar regions—typically are so remote that these moving masses of ice seldom creep into the human consciousness.

Yet glaciers, for all their remote mystery, play a crucial role in the nurturing of life on earth. By locking up some 75 per cent of the planet's fresh water, glaciers prevent a catastrophic rise in sea level that would inundate New York, Tokyo and other great coastal cities, changing the very shape of the continents themselves. And their colossal masses of ice, lying at the extremities of the earth, are important in the formation of global weather patterns.

Still, scientists see far more in the glaciers than just passive reservoirs of water or breeders of weather. In the 1980s, glaciers began taking on the quality of major natural resources. Meltwater from glaciers was such an important source for hydroelectric power and crop irrigation that scientists were examining the possibilities of growing glaciers artificially. Icebergs—the glaciers' gigantic offspring—were viewed as another potential source of fresh water that one day might be harnessed and towed to Saudi Arabia, California and other parched regions of the world. Still more important, perhaps, the greatest glacier of all—the ice sheet in Antarctica—had in recent years become an important natural laboratory for scientists whose findings promised not only to enhance man's knowledge but also to enrich the world's shrinking supply of food, minerals and other resources.

Yet of all the ways in which glaciers influence life on earth, the most significant is the power of ice to remake the very surface of the planet. This power, which the 19th Century Swiss naturalist and glaciologist Louis Agassiz called "God's great plough," was abundantly evident in the Peruvian avalanche: Enormous rocks were quarried, the mountainside was planed down, bedrock was scoured, vast mounds of gravel, sand and clay were deposited, a river was dammed and diverted from its channel. Into a mere seven minutes were telescoped processes that ordinarily take glaciers thousands of years to complete.

These actions by which the landscape is sculptured depend upon an essential element of the glacier's internal dynamics. Regardless of whether a glacier is advancing or retreating, the particles of ice within it are constantly flowing downslope. As the ice at the bottom or sides of a glacier moves, it freezes onto loose rock and soil, picking them up and carrying them forward. These materials, in turn, serve as the tools of the glacier. Sand and gravel carried by the ice abrade the landscape; boulders fixed in the ice gouge the land, leaving deep scars in the bedrock and prying loose large chunks to be incorporated into the glacier's cargo.

Only in recent years have scientists been able to observe directly the process of glacier erosion at work. By digging tunnels deep below the surface of a glacier, for example, they learned to measure the speed at which the bedrock was being worn down. In one experiment, researchers bolted a slab of fine-grained marble to the bedrock beneath a glacier in Iceland. For three months, the rock-studded bottom of the glacier flowed over the marble slab, moving a distance of 31 feet. Then the marble was removed. Its surface had been ground down by about a tenth of an inch —little enough, but when this is multiplied on the glacial timetable of thousands of years, the power of ice to transform the earth becomes dramatically explicit.

Indeed, much of today's landscape—jagged mountain peaks, deep-cut valleys, fertile plains, even lakes and rivers—is the legacy of glaciers that existed in the past on a much larger scale. About 20,000 years ago, for example, ice was remodeling fully one third of the earth's land surface—an area more than three times the extent of present-day glaciers. Vast ice sheets like those now in Greenland and Antarctica covered half of North America and large areas of Europe and Asia; smaller glaciers spread out from the Andes of South America and mountainous regions of Australia.

Though these great glaciers profoundly altered the face of the earth, their very existence—to say nothing of their dynamics—was a matter of great dispute as recently as 150 years ago. In the early 1800s, geology was still a science in swaddling clothes, and scarcely anyone studied glaciers. Most geologists regarded glaciers as esoteric and inconsequential phenomena peculiar to high altitudes or the high latitudes of the polar regions; no one dreamed that they played a considerable role in the grand design of the earth's visage.

The evidence left by ancient glaciers was all around, of course—grooved and polished bedrock, hills of stony debris called moraines, rounded rock outcrops known as *roches moutonnées*—loosely, muttoned rocks—because clumps of them vaguely resembled wigs slicked down with mutton tallow. Such features were attributed not to ice, however, but to the action of ancient water—the ever-popular, all-explanatory Biblical Flood that Noah had survived by building his ark.

Still, certain anomalies of the landscape puzzled the most ardent champions of the Flood doctrine. For example, the stony debris, which the early geologists called drift because it was thought to have been carried by water, was piled up in chaotic jumbles. Floodwater would have deposited the debris much more evenly, sorting it out by size and density, and leaving several layers, with heavy stones at the bottom, then gravel and sand.

21

Particularly perplexing were the boulders early geologists labeled erratics because they were found out of place—often at great distance from the bedrock where they obviously originated. (Common folk referred to these boulders as "foundlings" and suggested that witches had scattered them about.) Erratics of Scandinavian origin dotted the countryside of eastern England; it would have been necessary for the ancient Flood somehow to have carried them across the North Sea, which seemed highly unlikely.

In attempting to solve the problem presented by the erratics, Charles Lyell, a leading British geologist of the early 1800s, arrived at an imaginative solution. Lyell, in his classic textbook *Principles of Geology,* published in 1833, suggested that the erratics had been transported inside icebergs that had broken off from glaciers near the North Pole and then had been borne south by the waters of the Flood.

Though erroneous, Lyell's idea at least contained a germ of truth: Ice could in some fashion transport rock. Though this notion was a novel one for scientists, it was a long-established fact to the untutored farmers and others who lived near the glacier-studded Swiss Alps. These people perceived all too well that glaciers moved, for their ancestors had seen ice flows encroach upon their farms and settlements and then retreat. They also knew from their own observations that the moving ice left telltale grooves in exposed bedrock and carried and deposited quantities of rocky debris. By comparing these clues found near glaciers with scratched bedrock and rock piles farther away, they concluded that glaciers of the past must have been bigger—and responsible for the work mistakenly attributed to the Flood.

How these sensible observations emerged from the obscurity of peasant wisdom into the highly contentious councils of science is a fascinating tale. The chain of circumstance commenced in 1815. A Swiss mountaineer, Jean Pierre Perraudin, who made his living by hunting wild chamois in the Val de Bagnes, began talking to a highway and bridge engineer named Ignatz Venetz. He told Venetz about his idea: Though glaciers now occupied only a portion of the valley, they must have once filled the entire Val de Bagnes. Venetz did some field investigations and developed the theory that not just a valley in Switzerland but all of that country and other parts of Europe as well had once been covered by immense sheets of ice.

When Venetz presented his glacier theory at a meeting of the Swiss Society of Natural Sciences in 1829, however, only one man in the audience showed much interest. He was Jean de Charpentier, an amateur naturalist and director of the salt mines at Bex in western Switzerland. Curiously, de Charpentier had first heard the basis of the glacier theory 14 years earlier, from the same chamois hunter Perraudin, but had rejected it. Now the idea did not seem quite so farfetched; de Charpentier seized upon Venetz' version of Perraudin's theory and began organizing and classifying the evidence in support of it. But when de Charpentier presented his carefully systematized version to the same scientific society in 1834, he encountered no more enthusiasm than had Venetz. The dogma of the Flood—in its original form or in Charles Lyell's iceberg modification—was still solidly entrenched.

De Charpentier later remarked upon the irony of his icy reception by the assembled scientists. On the way to the meeting in Lucerne he had come across a woodcutter who had no trouble accounting for the presence nearby of a large granite boulder: It obviously had been deposited there by a glacier. The woodcutter, wrote de Charpentier, "was greatly astonished when he saw how pleased I was by his logical explanation, and when I gave him some money to drink to the memory of the ancient Grimsel glacier." The peasants accepted matter-of-factly a truth that the scientists refused to face.

Among the skeptics in de Charpentier's audience was Louis Agassiz, pro-

With a spectacular glacier visible through a nearby window, Swiss engineer Ignatz Venetz works on a map at his desk in this 1826 painting. Intrigued by the theory of a local hunter that striations on valley rocks were caused by glaciers, Venetz confirmed the idea with his own observations.

Perched precariously atop a pillar of ice that has been preserved by its shade, a huge boulder dwarfs visitors to Switzerland's Unteraar Glacier in this late-18th Century etching. The presence of such erratics—so called because they do not match rock formations nearby—led Venetz to theorize in 1829 that glaciers had once engulfed northern Europe, transporting tons of debris over long distances.

fessor of natural history at the College of Neuchâtel. An expert on fossil fish, Agassiz at the age of 27 already was considered one of Europe's foremost naturalists. He saw little merit in de Charpentier's theory of glaciers, but he had been a pupil of the speaker's, and he did not let scientific disagreement destroy an old friendship. Two years later, Agassiz accepted de Charpentier's invitation to spend the summer at Bex. There, Agassiz visited nearby glaciers and studied the moraines and erratics of the Rhone valley. The evidence was overwhelming, and to de Charpentier's delight, Agassiz quickly became a convert to the glacier theory.

The following summer, on July 24, 1837, the Swiss Society of Natural Sciences once again heard a discourse on glaciers. The speaker this time was Louis Agassiz, the new president of the society, and what the members heard was couched in terms that were far more dramatic than the previous presentations by Venetz and de Charpentier. Agassiz's quick mind had vaulted from the prosaic evidence of erratics and heaps of stones to a vision of an ancient time when the temperature plummeted temporarily throughout the world. He saw an "epoch of intense cold," when "Siberian winter established itself for a time over a world previously covered with a rich vegetation and peopled with large animals." He asserted that "death enveloped all nature in a shroud" as enormous glaciers descended from the North Pole and extended over much of the Northern Hemisphere, reaching as far south in Europe as the Mediterranean. (Later, borrowing a phrase coined by a colleague, he would refer to this glacial epoch as *Eiszeit*—Ice Age.)

Agassiz's dramatic phrases and scientific prestige got him scarcely any further with the Swiss society than the less-grandiose glacier presentations by Venetz and de Charpentier got them. His ice sheets sparked fiery debate, but Agassiz stood virtually alone. Even his mentor and patron, the prominent German naturalist Alexander von Humboldt, wrote him a few months later: "Your ice frightens me. I am afraid you spread your intellect over too many subjects at once." In effect, Humboldt was telling Agassiz to stick to his fossil fish.

The widespread resistance to the glacier theory stemmed in part from diehard adherence to the old Flood notion. But it also derived from ignorance about glaciers. Few scientists could even imagine ice sheets of the magnitude described by Agassiz; the immense Antarctic ice sheet had not yet been discovered, and it would be another 51 years before a scientific expedition established that the glaciers of Greenland formed a single large sheet. Moreover, the very extravagance of some of Agassiz's speculation worked against him. His gift for synthesizing ideas frequently led him to jump impulsively ahead of the evidence. For example, there were no deposits of drift in southern France to support his claim that the ancient ice sheet had reached the Mediterranean.

But Agassiz, once aroused, was as relentless as the ancient glaciers he had envisioned. He had the energy, intellect and inclination to juggle several important projects at once—glaciers *and* fossil fish *and* starfish and whatever else struck his fancy. He was possessed of what his father, a Protestant clergyman, described as a "mania for rushing full gallop into the future."

Perhaps most important, Agassiz was one of the first great organizers of modern science, adept at winning research grants from individual patrons and at building a staff to spend them. He lived modestly but spent extravagantly for science. His house in Neuchâtel was a combination laboratory and boardinghouse. There, during the 1830s, Agassiz assembled a 12-man research group that included his own artist and a fossil collector who, much to the dismay of Agassiz's wife, slept in his clothes and almost never changed them. In addition, Agassiz maintained a printing plant with no fewer than 20 employees who, when not at work publishing Agassiz's own voluminous research, were hired out to other authors.

Now, following the rejection of his glacier theory, Agassiz mustered his organization for a full-scale assault on the ice, aimed at proving his ideas. He even prevailed upon the King of Prussia to help foot the bill. (Though Neuchâtel was a Swiss canton, it was under nominal Prussian rule until 1857.) For nearly a decade, Agassiz and his entourage of assistants and students spent every summer in the Alps, climbing Mont Blanc, the Jungfrau and other peaks to investigate existing glaciers and search out evidence of ancient ones. Agassiz, now in his late thirties, looked so youthfully robust clambering up and down the mountains and glaciers that one traveler inquired if he was the son of the celebrated professor of Neuchâtel.

In the summer of 1840, Agassiz established a permanent research station on a rocky moraine next to the Unteraar Glacier, 40 miles southeast of Bern. What he dubbed the Hôtel des Neuchâtelois was in fact a crude stone hut. It consisted mostly of a huge erratic boulder with a jutting overhang that formed a roof; two walls were made of stones; the front opening closed with a blanket. The "hotel" could accommodate six sleepers elbow to elbow. Outside, a recess sheltered by another big rock served as kitchen and dining room. Ice from the Unteraar Glacier cooled the scientists' wine.

This became the base from which Agassiz waged his campaign in behalf of the glacier theory. (To see his busy father that summer, five-year-old Alexander Agassiz had to be carried up the mountain on the shoulders of a sturdy guide.) Here, Agassiz corrected the proofs for his book *Studies on Glaciers,* which incorporated findings from his field investigations. The book was handsomely printed by Agassiz's own shop and beautifully illustrated by his own artist. But its publication that autumn was overshadowed by the first of several nasty quarrels that occasionally marked the relationships between Agassiz and his closest colleagues. In this case, though Agassiz had dedicated the book to his predecessors in the glacier theory, Venetz and de Charpentier, its publication before the latter scientist could get into print with his own research led both men to break with Agassiz.

Oblivious to the carping of adversaries and former friends, Agassiz plunged on. Glaciology was a new science, and he had to devise his research techniques as he went along, improvising ways to pry the glacier's secrets from its impassive ice. To measure the thickness of the Unteraar Glacier,

Swiss scientist Louis Agassiz delivers a lecture in the United States in 1871. Although Agassiz was a naturalist by training—specializing in fossil fishes and marine biology—he became intrigued by glaciers as a young professor and spent a decade attempting to prove that they had once covered much of the earth's surface.

Agassiz and his colleagues had to strap iron rods to their backs and then lug them onto the ice. The rods were attached to one another and hammered into the ice, first 100 feet, then 200 feet. "As well might I have tried to sound the ocean with a ten-fathom line," wrote Agassiz. Only after repeated trips up the mountains and the assembly of 1,000 feet of rod did he succeed in reaching the bottom of the glacier.

On another occasion, Agassiz decided he had to plumb personally the frozen heart of the Unteraar Glacier. For his descent he chose one of the so-called moulins—deep vertical shafts in the glacier caused by the melting that occurs when streams encounter fractures at the surface and erode them further. A tripod was rigged over the moulin, and a board swing was suspended from it. Agassiz, seated on the board, was then lowered into the depths of the glacier.

At first the going was easy. Agassiz twirled slowly as he passed into the glittering world of ice. His friend and fellow naturalist Arnold Escher von der Linth lay prone at the edge of the moulin, relaying directions between Agassiz and the crew of men shouldering the rope. Should anything go wrong, Escher could have the board swing hoisted back up.

Below, Agassiz turned his attention to his first scientific observation. He and other scientists had noticed that as far as they could see down into any of these moulins the glacier seemed to be composed of alternating blue and white bands; this indicated that the structure of the ice in the glacier was laminar, that is, lying in layers. Agassiz suspected that these laminations were somehow related to the glacier's movement and he was intent upon being the first to observe exactly to what depth they reached.

He was now 80 feet into the hole, dangling in a surreal world of sculpted ice suffused with a glowing light. Above him Escher's head grew smaller and smaller against the bright circle of sky at the mouth of the moulin. At this depth the shaft divided into two parts like the legs of a pair of trousers. Agassiz chose the wider of the two ways but soon found it repartitioned into a number of impassable holes. Upon signaling to Escher, he was hoisted back up to the bifurcation and then allowed to descend the other passage. He was now more than 100 feet below the surface of the glacier.

The blue bands of ice still sparkled before him in mesmerizing patterns, and Agassiz was momentarily lost in contemplation of them when his feet suddenly plunged into ice-cold meltwater; he was near the floor of the glacier. This was no great predicament—but his outcry was misinterpreted by his friends above, who continued to pay out more rope. Agassiz sank helplessly into the water, and only after he had sent aloft repeated shouts of distress did the rope tighten. But now Agassiz found himself drawn upward to even greater danger.

Lining the walls of the moulin were hundreds of pendulous ice stalactites. Needle-sharp spears of ice formed by the continuous freezing of dripping water, these stalactites had presented no danger on the trip down, but now as he was pulled upward he was in constant danger of being drawn onto one of their points or, even worse, being skewered by one that had been broken off by the slapping of the rope above him. These loose arrows of ice came rattling down the shaft as Agassiz nimbly ducked and steered his way upward. At last, after a harrowing ascent, he was drawn into the arms of his friends. "I should not," he remarked shortly afterward, "advise anyone to follow my example."

Agassiz's research base, the "hotel," quickly became a gathering place for skeptical geologists from all over Europe. They wanted to see for themselves the evidence for his glacier theory. Among the visitors was the Reverend

A medial moraine of rocky debris marking the confluence of two large glaciers in the Swiss Alps provides an observation platform for glaciologists working with Louis Agassiz in 1840. Near a prominent boulder at right, two weary observers rest in a crude field station.

William Buckland, a leading British geologist. Buckland was considered one of the most colorful professors at Oxford, a university noted for its eccentrics. He wore his academic robe and a top hat even when he ventured into the field. He constantly carried a blue bag of fossils from which, wrote one contemporary, "even at fashionable evening parties, he would bring out and describe with infinite drollery the latest 'find' from a bone cave."

Buckland was also one of England's staunchest proponents of the Flood doctrine. His visit to Agassiz's Alpine station during the summer of 1840 shook his faith in the Flood, but he was still not converted to glaciers. During that autumn, however, Buckland invited Agassiz on a series of field trips through northern England and the Scottish highlands. Buckland wanted his guest to see Scotland's famed Glen Roy, a lovely valley nearly surrounded by a puzzling series of terraces. These terraces looked like parallel roads carved from the walls of the valley at different heights. There were many theories about what caused the terraces, but the Flood figured in most of them. When Agassiz provided a clear explanation for the puzzle—they were terraces left by the gradual recession of a lake formed when ancient glaciers dammed up the valley—Buckland was convinced. Now a zealous convert to the glacier theory, Buckland helped convince Charles Lyell and other influential English geologists. A few months after Agassiz's visit, a colleague wrote him from England: "You have made all the geologists glacier-mad here, and they are turning Great Britain into an ice-house."

With important opinion beginning to swing his way all over Europe, Agassiz completed a second book, *Glacier Systems,* in which he chronicled the Alpine glaciers and discussed the genesis of the ice sheets that he theorized had preceded the uplift of the Alps. He then set his sights on a new world to conquer. In 1846, with $3,000 from the King of Prussia and an invitation from one of the influential Lowells of Boston to lecture there, he set sail for America. When his ship docked briefly at Halifax in Nova Scotia, Agassiz wrote later, he "sprang on shore and started at a brisk pace for the heights above the landing. On the first undisturbed ground, after leaving the town, I was met by the familiar signs, the polished surfaces, the furrows and striations, so well known in the Old World; and I became convinced of what I had already anticipated as the logical consequence of my previous investigations, that here also this great agent had been at work."

Both Agassiz's glacier theory and the man himself were a great success in Boston. The next year he accepted a specially created professorship at Harvard University and settled permanently in Cambridge. He quickly became one of America's best-known scientists, hobnobbing with such literary greats as Ralph Waldo Emerson and Henry Wadsworth Longfellow. As a teacher, he influenced students such as William James, who later became America's leading psychologist, and social historian Henry Adams, who wrote that Agassiz's course on the ancient glaciers had more effect on his curiosity "than the rest of the college instruction altogether."

In America, perhaps because his beloved Unteraar Glacier was so far away, Agassiz's interest in ice ebbed and his research turned again to his first love—zoology and natural history. In fact, during the last decade of his life, he devoted considerable time to attacking the new theory of evolution proposed by Charles Darwin. Ironically, Darwin had been one of the first scientists in England to accept Agassiz's radical ideas about the ancient glaciers. But Agassiz remained an unswerving creationist. The ice sheets had destroyed all life, he insisted, and thus there could be no connection between the species of past and present as Darwin maintained; the Creator must have made life all over again after the Ice Age.

Agassiz's obsession against evolution apparently led him, during an expedition to Brazil in 1865, to misinterpret weathered rocks in the Amazon valley as evidence of widespread glaciation there. In his view, the greater the size of the ancient glaciers the less the chance that any life had survived, and if nothing had survived, then Darwin had to be wrong. By then, writes his biographer Edward Lurie, Agassiz was attempting to combat Darwinism by "covering the whole world with ice."

After Agassiz's death in 1873 at the age of 66, a special monument was brought from Switzerland to mark his grave in Boston: a 2,500-ton erratic boulder, scarred and polished by ice and found near the site of his old Hôtel des Neuchâtelois next to the Unteraar Glacier.

Equally fitting as a memorial was the fact that by the time of his death the glacier theory, so roundly scorned when he had presented it 36 years before, was firmly established on both sides of the Atlantic. While Agassiz's own interest in ice had waned, hundreds of other scientists had become infected by his early enthusiasm and were swarming over glaciers and sifting through the stony debris left by the old ice. These investigators found evidence that glaciers had overwhelmed large parts of the earth not just once, as Agassiz had believed, but several times. Deposits of glacial drift sometimes contained several distinct layers of rocky debris separated by layers of peat and other decomposed vegetation. This meant that, after the retreat of each glacier, a warmer climate had prevailed, enabling vegetation to flourish before ice once again enveloped the land.

A century of research since Agassiz's death has established that glaciers overran large portions of the earth on at least four and perhaps on as many as 10 occasions during the last million years, which constitute the geological epoch known as the Pleistocene. These periodic invasions of ice during the Pleistocene occurred at approximately the same times around the world. They followed the same general routes of expansion on each occasion and covered about the same total geographical area—roughly 17 million square miles.

Contrary to Agassiz's supposition, however, the ice did not begin at the North Pole and then move south. As temperatures dropped (for reasons that are still not clear), small glaciers in upland areas all over the world gradually enlarged. These streams of ice flowed down into the valleys where they joined forces and eventually merged into sheets of ice up to two miles thick. Then, like pancake batter on a griddle, these sheets spread in all directions, overwhelming the landscape.

In North America, the largest sheet, the Laurentide, spread out from near what is now Hudson Bay. It covered more than five million square miles, reaching north to the shores of the Arctic Ocean and south to bury all of eastern Canada, New England and much of the northern half of the midwestern United States. A smaller sheet, originating in the Canadian Rocky Mountains, engulfed western Canada, parts of Alaska and small portions of the northwestern United States.

In Europe, an ice sheet radiated from northern Scandinavia to cover most of Great Britain and large parts of northern Germany, Poland and the Soviet Union. A smaller sheet, centered in the Swiss Alps, covered parts of Austria, Italy, France and southern Germany. Asian ice sheets occupied the Himalayas and parts of Siberia.

The Southern Hemisphere—with the obvious exception of the Antarctic ice sheet, which had nowhere to go except the sea—experienced significantly less glaciation during the Pleistocene. Small ice sheets expanded in Australia, New Zealand and the Andes of South America.

These periodic invasions of the past million years are only the tip of the ancient ice. Geologists have recently found scarred bedrock, striated pebbles and other traces of glaciation that date back 700 million years or more. Evidence has been uncovered even in such unlikely places as Africa's Sahara. There apparently have been many ice ages, stretching back toward the dim beginnings of the planet, and it is likely that practically every country on earth bears some marks left by ancient glaciers.

For the most part, however, the effects of these earliest glaciers on the landscape have been obscured by millions of years of erosion, mountain building and other geological change, including succeeding glaciations. It was the more recent Pleistocene glaciers that put the finishing sculptural touches on much of the earth. The last of these glacial invasions—known as the Wisconsin Stage to scientists in the United States and by other names elsewhere—began about 70,000 years ago, reached its maximum 20,000 years ago and ended 8,000 years ago when the glaciers shrank to approximately their present-day limits.

This precise dating of the Wisconsin Stage glaciers had to await modern methods of analysis, such as measuring the amount of radioactive carbon in organic materials found among the drift deposited by the ice. But even during the latter part of the 19th Century, long before such sophisticated dating techniques became available, sufficient clues were present for the enthusiastic heirs of Louis Agassiz—geologists, geographers and glaciologists—to chart the extent of the most recent ice invasions. The clues were the multitudinous landforms—from scratched rocks to jagged mountain peaks and fertile plains—that are the legacy of, in Agassiz's felicitous phrase, "God's great plough."

Perhaps the most spectacular features of ice-sculptured terrain can be found in the mountains where most glaciers originate. Snow accumulates in small hollows on mountainsides, and over many years it changes into ice by recrystallization—refreezing of the meltwater and compression; eventually the ice may become thick enough to move because of its own weight. Bedrock lying at the head of the newborn glacier, and under it, already has been jointed or fractured by centuries of frost action—the alternate expansion of water through freezing (about 9 per cent by volume) and then contraction through thawing. The glacier's icy fingers probe into the fractures and adhere to the rock fragments, which are then carried along by the moving mass of ice. The rocks, now frozen in the bottom layers of the glacier, pry loose further fragments, which also are incorporated into the ice. Thus, the floor, headwall and sides of the hollow are gradually scooped away. The result is a circular steep-sided basin that geologists call a cirque.

Cirques sometimes grow quite large—up to a mile or more in diameter, with walls several hundred yards high. If cirques scallop all sides of a mountain, they may bite into it so deeply that only a single jagged peak, or horn, remains. The classic example is Switzerland's Matterhorn, but thousands of other peaks are glaciated horns, including the highest of them all, the 29,028-foot Mount Everest.

By creating horns, glaciers have presented mountain climbers with their greatest challenge. But if glaciers have provided the problem they also have occasionally offered the solution. In 1953, when Sir Edmund Hillary and his Sherpa guide became the first climbers ever to reach the summit of Everest, the ascent was achieved in part by traversing the Khumbu Glacier, which thousands of years ago began to carve out the huge cirque on the mountain's southwest face.

The second act in the ancient drama of glacial erosion occurred when the streams of ice descended from the mountains into the plains and major river valleys, merging into larger valley glaciers and enormous sheets. The thicker the ice the more effective the glacier as an engine of erosion. A glacier only half a mile thick exerts pressure of 62 tons per square foot. This pressure crushes all the material beneath and adjacent to it and pushes the glacier's rasplike tools—the jagged rock fragments frozen into the ice—deeper into the underlying bedrock. In parts of Canada today, the airborne viewer can look down upon the tracks of the ice sheet—great furrows in the crystalline bedrock more than 20 yards deep and a mile long.

Erosion by ice is also influenced by factors other than the thickness of the glacier. Soft bedrock such as limestone crumbles fastest, and hard rocks like granite and quartzite make the most effective cutting tools in the ice's grip. The glacier's erosive power also depends upon the quantity of rocks it carries and upon its speed of movement.

When the ancient ice was squeezed into narrow river valleys, it speeded up. It tore at the floor and slopes, changing the V-shaped profile with gentle slopes characteristic of stream erosion into deep U-shaped troughs. A superb example is California's Yosemite Valley, which experienced the widespread glaciation—though not a full-scale ice sheet—that affected the Sierra Nevada about 20,000 years ago.

Decked out in a traveling cloak and carrying a roll of maps, British geologist William Buckland is gently lampooned in this contemporary cartoon. An early opponent of the new glacial theories, Buckland abruptly changed his views after observing evidence of glaciation in Scotland in 1840.

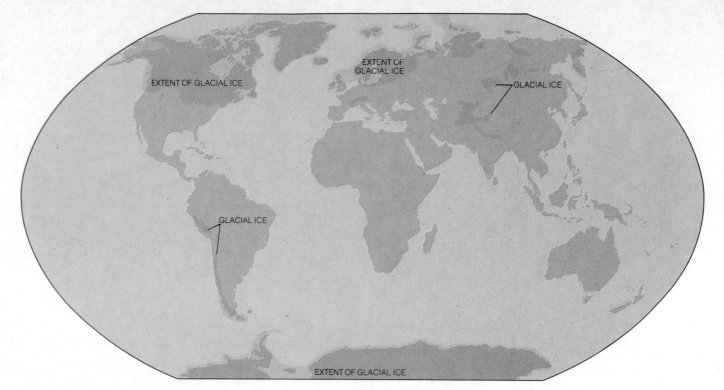

EXTENT OF GLACIAL ICE

EXTENT OF GLACIAL ICE

GLACIAL ICE

GLACIAL ICE

EXTENT OF GLACIAL ICE

Colossal ice sheets smother much of the Northern Hemisphere in this map of the globe as it appeared 20,000 years ago. Glacial ice one mile thick covered as much as 23.5 million square miles, locking up so much of the planet's water that sea levels fell 350 feet.

During the late 19th Century, the American naturalist John Muir was one of the first investigators to attribute the shape of Yosemite Valley to glacial erosion (on hands and knees he studied the telltale bedrock scratches with a magnifying glass). Muir liked to tell of the time in 1879 when he led a horseback tour through Yosemite. One of the tour members, a noted and rather corpulent clergyman, disputed Muir's view about the glacial origin of the valley; Yosemite had come into being all at once, insisted the clergyman, "created out of the hand of God." When Muir pointed out a section of glacier-polished granite, the clergyman dismounted to inspect it and his iron-nailed shoes slipped on the slick rock surface, sending him sprawling. "Now, Doctor," said Muir, helping him to his feet, "you see the Good Lord has given you this most convincing proof of the mighty work the glaciers have wrought!"

This mighty work is only partially visible to the 2.6 million tourists who visit Yosemite in a year. They can see the spectacular Yosemite Falls that plunge a total of 2,425 feet to the valley floor from a hanging valley carved out by a tributary glacier that did not cut as deeply as the main stream of ice. But scarcely more than half of the height of El Capitan and the other sheer granite cliffs that hem in Yosemite Valley are visible. Seismic soundings have shown that the glacier gouged out a trough up to 2,000 feet deeper than the present height of these walls, but this space was filled by clay and other sediments left by a lake that occupied the valley after the retreat of the ice.

In other places, bodies of water help obscure the extent to which ice scoured old river valleys. The Finger Lakes of upstate New York occupy U-shaped valleys deepened by the ice sheet. The Finger Lakes (named by early settlers in the belief that the 10 lakes represented the imprint of the Creator's hands) point to the south—the direction of the glacier's movement through their basins. Scotland's Loch Ness, home of the fabled but elusive monster, is also a U-shaped valley gouged out by glaciers.

Glacier-carved valleys that open onto the sea are fjords, which are typically long and narrow with steep cliffs. These inlets indent the coastlines of many regions from southern Chile to Greenland, which is surrounded by them. But fjords are most associated with the Atlantic coast of Norway, where their deep channels and protective walls long have provided safe anchorage for seafarers. The Sogne Fjord near Bergen is 120 miles long and has sides that tower up to 4,000 feet above the water and extend 4,000 feet below it. An ancient glacier nearly a mile thick descended from the mountains east of there, plowed into the sea and continued to cut a U-shaped trough in the sea bottom until the water was deep enough—about 90 per cent of the ice's thickness—to make the ice float.

An ominous characteristic of some fjords and other U-shaped valleys is their steep walls. In regions where the erosion is relatively recent by glacial standards—perhaps 10,000 years old—the freezing and thawing action of frost can jar loose rock fragments that give birth to landslides. In the narrow confines of Norwegian fjords, the destructive effects can be intensified when landslides crash into the water. Norway's Innvick Fjord has been the scene of at least two tidal waves set off by rockslides; on one occasion, a steamship was lifted and swept more than a quarter mile overland.

But the power of moving ice is infinitely greater than that of moving water. The ancient glaciers, in addition to carving fjords like Norway's Innvick, U-shaped valleys like Yosemite, and spectacular mountain cirques and horns, also crept inexorably across plateaus and plains areas, stripping away the covering of topsoil and surface bedrock. In Canada alone, it has been estimated that the glaciers removed an average depth of 30 feet of rock over a region of two million square miles.

Like a conveyor belt, the flowing ice carried along all of this drift—from enormous boulders to finely pulverized particles known as rock flour. (Everything carried by glaciers is still called drift, from those early days when geologists maintained that glacial debris actually had been "drifted in" by the Biblical Flood.) When any part of the ice melted, drift was dumped onto the landscape, further remodeling it. The rocky detritus that had served as the glacier's tools of erosion now piled up to create a variety of new landforms.

The drift that makes up these new landforms is classified according to the two principal ways it can be deposited. Material deposited directly by melting ice—a conglomeration of unsorted clay and stony debris—is known as till; the name derives from the Scottish word meaning "obstinate land," and is appropriate because most farming in Scotland takes place on a thin layer of soil that overlies obstinate glacial rubble. The second kind of drift, which has been neatly sorted by the action of running water, is called outwash because it was carried away from the glacier by meltwater.

One distinctive landform consisting of till is the drumlin, an elongated hill shaped like the back of a whale. Drumlins typically are about a mile long, perhaps a third of a mile wide and up to 200 feet high. Drumlins were plastered on the landscape as ice melted along the bottom of the glaciers, then were molded and streamlined by the pressure of the ice passing over them. Whenever fields of drumlins are found—for example, in Ireland, where the name (meaning "small ridge") originated, or in upstate New York, where local people call them "hoddy-doddies"—they are always aligned in the direction of the ice's movement. The most famous drumlin is Bunker Hill in Massachusetts (unjustly so, since the Battle of Bunker Hill actually was fought on an adjacent drumlin, Breed's Hill).

Another distinctive landform consisting of till is the end moraine, a ridge or hill up to several hundred feet high that was created by the melting of the end of a glacier when the glacier's terminus remained at the same place for a considerable period. One type of end moraine, the terminal moraine, is of particular interest to the geologist because it marks the farthest point of

V-SHAPED VALLEYS

Before a glacial onslaught, a mountain landscape displays the gentle contours of gradual erosion by wind and rain. Streams have cut a typical network of V-shaped valleys through the soil that blankets the hillsides.

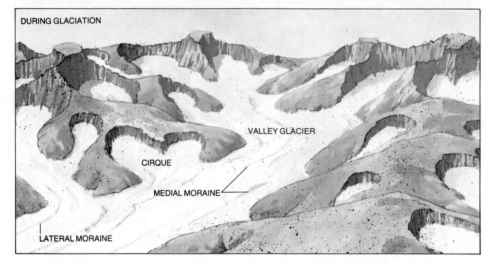

DURING GLACIATION

VALLEY GLACIER

CIRQUE

MEDIAL MORAINE

LATERAL MORAINE

The glacial transformation of the land begins in hollows where gradually accumulating ice carves out deep, bowl-shaped cirques as it begins its downhill flow. The glaciers scour soil and rock from the lower slopes, redepositing the debris in gracefully curving moraines.

AFTER GLACIATION

HORN

ARÊTE

U-SHAPED VALLEYS

CIRQUE LAKE

HANGING VALLEY

The retreat of the ice leaves a rugged tableau, dominated by ice-carved peaks called horns and sharp, ragged ridges known as arêtes. Dramatic hanging valleys carved by tributary glaciers join the deep U-shaped valleys that mark the wake of the glacial mainstream.

advance by the glacier. Terminal moraines in the United States and in northern Europe typically are curved, suggesting that broad lobes jutted out to form the front of the main part of the ice sheet. In the 19th Century, before the building of large American cities and the clearing of land for farming obliterated many of these terminal moraines, geologists were able to trace a more-or-less continuous series of ridges from eastern Long Island to the state of Washington, delineating the southern limits of the ancient ice sheets.

The ice sheets created many end moraines, of course. Every time the glaciers paused in their retreat a new ridge or hill of rocky debris was deposited. In addition, melting ice at the bottom of the glaciers deposited great sheets of till called ground moraines. Vast regions of the American Middle West and of northern Europe consist of ground moraine, up to several hundred yards thick, that buried preglacial hills and valleys. Most of the peninsula occupied by Denmark consists of jumbled-together ground and end moraines built up by successive advances of the ice sheets—to a depth of more than 365 yards in places.

The distinction between unsorted till deposited directly by a glacier and the stratified outwash deposited by meltwater is nicely illustrated by the beaches of New York's Long Island. The northern length of the island contains terminal moraines, and the bouldery beaches of the north shore reflect these till deposits. The southern length of the island, by contrast, is made up of outwash plains—sand and gravel carried away by glacier meltwater and deposited in distinct layers. Along the south shore, this outwash has been further worked by the waves of the Atlantic, creating magnificent beaches of fine-grained sand.

In the instance of Long Island, the outwash plains lie beyond the edge of the glacier's advance. But other landforms of outwash with quaint names such as eskers and kames were deposited near or under the glaciers. Kames are small mounds or terraces left by meltwater flowing over the surface of glaciers. Eskers are winding, serpent-like ridges up to a few hundred yards long that resulted from deposits left by streams of meltwater flowing in tunnels at the bottom of the ice. In kames and eskers, sand and gravel often are so uniformly graded that they look as if they have passed through a sieve. Both landforms thus are valuable sources of aggregate used in construction.

Among the most conspicuous constituents of drift, whether deposited as unsorted till or as nearly stratified outwash, are the erratics that puzzled early-19th Century geologists. Though erratics are ordinarily thought of as boulders, they can be rock of any size found out of its normal geological context. Some erratics are truly enormous—dramatic testimony to the power of ice to move everything in its path. One of Europe's best-known erratics, Pierre à Bôt (toadstone), weighs an estimated 3,000 tons and squats in the Jura Mountains of Switzerland, where it was borne by ice on a 70-mile journey from Mont Blanc to the south. Even bigger erratics—several of them slabs more than a mile long—have been found in Germany, Canada and the United States; in England, an erratic slab was the base on which an entire village was built at Huntingdonshire, about 50 miles north of London.

Smaller erratics served as sturdy building materials in both North America and Europe: They are the fieldstones from which fences, houses and even several castles in Germany were erected. By tracing erratics to their preglacial sources—the greatest distance on record is 800 miles—geologists can track the path of the ice that carried them. To the trained eye, as University of Wisconsin Professor Gwen Schultz points out, "a boulder of Precambrian crystalline rock from Ontario set down on a limestone ridge in Ohio is as conspicuously out of place as a polar bear on a Cleveland sidewalk."

But erratics can also be extremely small—and even more valuable to the finder. In eastern Finland, erratic rock fragments bearing copper ore led to the discovery of the bedrock source that became the nation's most important copper deposit. During the second half of the 19th Century, prospectors discovered modest amounts of gold among the terminal moraines and deposits of outwash that marked the glacier's farthest point of advance in southern Indiana.

Greater interest attended the small diamonds occasionally found by the prospectors who looked for gold in the glacial drift of Indiana. Diamonds

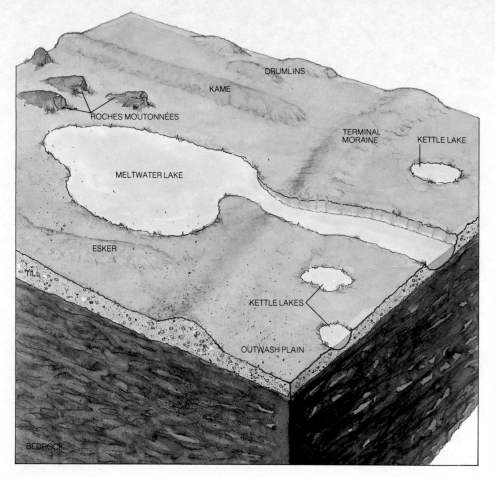

Certain landscape features are clear evidence of glaciation. Advancing over bedrock, the glacier sculpts rock outcrops into rounded hillocks called *roches moutonnées,* while debris called till deposited by previous glaciation is molded into elliptic drumlins under the ice. Loose till settling in a cavity builds a kame, and till in a meltwater tunnel beneath the ice creates a curved mound called an esker. A massive wall of debris, the terminal moraine, marks the end of the glacier's advance, damming meltwater to create a lake. As the glacier retreats, debris carried by meltwater streams builds an outwash plain where kettle lakes are formed by the melting of subsurface ice.

up to 21 carats in weight were also found during that period in the moraines and outwash of Wisconsin, Michigan and Ohio.

For a time, geologists entertained visions of finding the rich mother lode of these diamonds somewhere to the north. The search, following the course of the great ice sheets, took them into the Canadian wilderness between Lake Superior and Hudson Bay. Here, among the igneous rock known as kimberlite, they were certain, lay the deep vertical pipes—former conduits from within the earth's mantle—where diamonds were formed under enormous heat and pressure. Geologists never found the pipes, and finally quit looking. But to this day amateur rock-hunting enthusiasts in the American Midwest keep a sharp lookout for shiny erratics that might have found their way into gravel driveways and parking lots.

Particularly in the states of Wisconsin and Minnesota, the drift itself became the matrix in which the ice sheets molded yet another landform. Huge chunks of ice sometimes were left isolated under heaps of drift in front of the retreating glaciers. As the chunks of ice slowly melted, drift caved in around them, creating gently sloping depressions up to a mile in diameter that are known, because of their shape, as kettle holes or just plain kettles. By the tens of thousands, these kettles filled with water and turned into ponds and small lakes.

In fact, many of the world's lakes, small and large, are the products of one or more kinds of glacial action. In some instances, the ice sheets followed preglacial valleys and scooped out broad basins that became lakes. Sometimes the drift deposited by the ice served as dams to hold in the water. These two mechanisms—erosion and deposition—were responsible for some of the largest lakes in the world, including the Great Lakes of North America, the outlines of which suggest in some places the shape of the lobes at the edge of the ancient ice sheets. The combined forces of

The Tenaya Canyon in Yosemite National Park displays the breathtaking legacy of a glacial epoch that ended some 10,000 years ago. Granite walls have been carved into sheer cliffs soaring 2,400 feet above the valley floor, where fertile glacial till supports a dense forest.

erosion and deposition also had a profound effect on the patterns of river drainage. Old rivers were obliterated and several of North America's greatest rivers, including the St. Lawrence and Mississippi, were rerouted.

The multitudinous ways in which the ancient glaciers remodeled the landscape through erosion and deposition had been envisioned by Louis Agassiz, though not in the detail mapped by later geologists. After his death, however, it became clear that the ice had altered the landscape through yet other mechanisms.

One mechanism—the lowering of the oceans worldwide—was only temporary, but it had far-reaching ramifications. The moisture that, in the form of ice, nourishes glaciers comes ultimately through evaporation from the oceans. The growth of the great ice sheets consumed so much moisture that sea level fell by about 350 feet. Nineteenth Century geologists arrived at this figure not long after Agassiz's death by estimating the extent and the average thickness of the ice sheets. The terminal moraines laid down by the ice gave them the extent of glacier coverage; scarred bedrock and other evidence of glaciation on the sides of mountains told them how high the ice had reached. Their rough calculations have generally been confirmed by modern investigators.

The drastic lowering of the seas exposed land that long had been submerged. The floor of the English Channel dried up, uniting the British Isles with the European continent. In the Bering Strait, the 56-mile-wide strip of water separating Asia from North America, a broad isthmus appeared perhaps 20,000 years ago at the apex of the most recent period of glaciation. Over this bridge of land, which itself was not covered by ice, certain plants and animals—and most likely, human beings—made their first entry into the New World.

Meanwhile, just as the level of the sea was drastically affected by the

A Greenland glacier sculpts a new feature on the face of the earth. Its terminus, shown here, is slowly bulldozing rocky debris in its path into a ridge known as a push moraine.

growth and melting of the glaciers, so too was the level of the land. The ice sheets were so heavy that they depressed the crust of the earth. The pressure of the ice pushed the crust down into the denser heavy rock underlying the earth's crust, forcing this plastic material to flow outward—just as the addition of a passenger to a rowboat pushes aside water and makes the boat ride lower in the lake. In areas where the glaciers were thickest, the land sank by 2,000 feet or more, or about one third the thickness of the ice.

When the ice sheets receded and the weight lifted, the plastic mantle began to flow back. The land started to rebound, rapidly at first and then more slowly. The process of uplift is still under way in Scandinavia and the northeastern region of North America and will continue until the earth's crust reaches its preglacial state of equilibrium. The region around Hudson Bay, where the larger North American ice sheet began, has recovered only about half of the estimated 2,000 feet of elevation lost to the ice. At the current rate of uplift—about two feet per century—it will be nearly in equilibrium in some 5,000 years, leaving most of the bay high and dry.

Nowhere are the consequences of postglacial uplift more evident than in a small group of islands in the Gulf of Bothnia midway up the western coast of Finland. These islands are known as the Replot Skerry Guard; in Scandinavian languages, a skerry is a rocky island or reef, and any group of islands fringing the coast is called a skerry guard. The lives of the 3,000 or so people who inhabit the Replot Skerry Guard are influenced in many ways by the legacies of the last Scandinavian ice sheet, which retreated from the islands about 10,000 years ago. The people live on gravelly moraines and build their fences and houses from ancient glacial erratics. Most significantly, their rocky terrain is rising at the rate of approximately three feet every 100 years. Because the shallow sea floor around and between the islands is rising at the same rate, large sections of the sea floor are becoming dry land. In less than 200 years, the land area of the Replot Skerry Guard increased by 35 per cent, from about 69 square miles to 92. The two major islands, Replot and Vallgrund, became one, and scores of smaller islands emerged from the sea.

For generations now, the property owners of the Replot Skerry Guard have had the happy duty of meeting every 50 years to divide up the newly emerged land. Actually, the islanders are involved in a kind of trade with the sea. Fishing was once their principal industry, but now that the spawning grounds and harbors are drying up, the islanders are turning increasingly to agriculture, converting their new gift from the sea into pastures and cultivated fields that grow hay to feed their cows.

In fact, of all the many legacies bequeathed by the ancient glaciers—the beauty of scalloped mountains and deep-cut valleys, the utility of great lakes and rivers—the world's rich farmlands may be the ice's greatest gift. Louis Agassiz was one of the first to recognize the glaciers' boon to agriculture. He saw that the relentless action of the glaciers ground up the granite and the limestone and mixed them so that, as he put it, "the elements of the soil were mingled in fair proportions."

What Agassiz did not realize was that much of this enriched soil came to be distributed over the planet's most fertile plains by an indirect but most effective method. The particles of rock and silt transported by the glaciers were first carried away by meltwater, which sorted them out from coarser materials and deposited them in the plains of outwash beyond the melting ice. There, the particles dried and then were swept up by the fierce winds that howled off the ice sheets. The dust, blowing in great voluminous clouds, finally came to earth and stayed wherever there was vegetation to help anchor it.

This gritty wind-blown product is called loess, from an old German word meaning "loose" or "light." Not all loess is of glacial origin—in China, it derived from the sands of the Gobi Desert. But loess from the ancient ice covers an estimated one million square miles in North America and Europe, where loess deposits are sometimes more than 250 feet deep. In the United States, a thick blanket of loess forms the foundation for the midlands region, an area of almost two million square miles in the Mississippi River basin. Here, prairie grasses that helped to anchor the wind-blown dust also rooted easily between the loose particles of loess. After the ice sheets departed, the growth and decay of these grasses over thousands of years—combined with the mineral richness, good drainage and ease of cultivation of the loess base—built up the immensely fertile soil that helps feed the United States and the world.

Thus, every spring when the farmer in Iowa begins to prepare his fields for corn or soybeans, ice and glaciers may be the furthest things from his mind, but he is working land that already has been cultivated by "God's great plough." Ω

One of the most avid early students of glaciers was impelled by artistic, not scientific, considerations. Vittorio Sella, born in 1859 into a large family dominant in the textile industry in Biella, Italy, might easily have become a prosperous and forgettable businessman. Instead he was infected at an early age by two passions—his father's love for the emerging art of photography and his uncle's enthusiasm for mountaineering.

At the age of 19, Sella borrowed a bulky camera from a local photographer and lugged it to the top of Mount Mars near Biella to capture his first high-altitude panorama. Entranced by the results, he set about scaling other peaks, always carrying his cumbersome photographic equipment with him.

By the end of the 1880s, Sella had climbed most of the major peaks in the Alps, and his photographs of this icy realm were acclaimed throughout Europe. In 1889 he undertook the first of three journeys into the Caucasus Mountains in Russia; for his exhaustive documentation of the topography of this range, he was decorated by the Russian imperial government.

Sella soon found that he and his camera were in demand for expeditions with Italian royalty and with Europe's mountain-climbing elite. By the turn of the century, he had hosted an Alpine tour for Queen Margherita of Italy and had traveled to the Himalayas and to Alaska with Italy's Duke of Abruzzi. In company with the Duke, Sella closed out his mountaineering career in 1909 with a bold—but unsuccessful— attempt to scale the 28,250-foot Himalayan peak known as K-2. That same year, the 50-year-old photographer retired to Biella. There he remained until his death in 1943, surrounded by a visual record of glacier grandeur—sampled here and on the following pages—that has never been excelled.

Dwarfed by a colossal field of ice, a companion of Sella's *(lower right)* marvels in 1881 at great cavities that were carved by meltwater in the Aletsch Glacier in Switzerland.

Marching in stately procession, Queen
Margherita of Italy *(second from left)* and her
companions climb Monte Rosa in the
Pennine Alps in 1893. The intrepid monarch
made the 14,500-foot ascent to open a
high-altitude meteorological station.

High in the Karakoram Mountains near the border between India and China, the Duke of Abruzzi *(right)* and two guides scale the Chogolisa Glacier in 1909. Shortly after this photograph was taken, the expedition climbed to 24,000 feet on nearby Bride Peak— an altitude record that stood for 15 years.

Herdsmen chat outside their weathered huts overlooking the sprawling Forno Glacier on the craggy slopes of Italy's Valtellina Valley.

Two explorers survey the rugged surface of the Karagom Glacier in the central Caucasus Mountains in 1890. "Your hand shakes not only from the cold," Sella wrote of his difficult art of Alpine photography, "you fear a mistake, which would render useless all the discomforts and preparation."

Chapter 2

THE ANIMATED LIFE OF GLACIERS

Taken from the same spot, three photographs (*left and above*) shot in 1941, 1950 and 1967 record the swift retreat—a little over three miles in 26 years—of the Dawes Glacier along Endicott Arm in Alaska. The retreat began more than 50 years before the first picture was taken, but movement was slight during that period—only one mile—because the rocks at the bend where the fjord narrows trapped the glacier's snout. Once the ice retreated past the constriction, it broke up more rapidly, leaving the fjord choked with ice fragments.

On a fine fall day in October 1933, naturalists Matthew E. Beatty and Charles A. Harwell were making a routine survey of the Lyell Glacier in California's Yosemite National Park when they encountered a seeming impossibility. "I chanced to glance over to my right," wrote Beatty later, "where to my great astonishment I saw what appeared to be, at first glance, a normal living mountain sheep staring at us across the ice. I could scarcely believe my eyes."

There was every reason for amazement: Mountain sheep had been extinct in the Yosemite region for at least half a century. It hardly lessened the men's surprise to discover—as they almost instantly did—that the animal confronting them was not alive, but long since dead, and in the final stages of emergence from its glacial tomb.

As later studies would show, the sheep had perished perhaps 250 years before. It had fallen, or had been carried by an avalanche, off the face of the mountain above the head of the Lyell Glacier and into the bergschrund—the chasm between stationary ice and the glacier. The great stream of ice had encased the animal and had borne it along until at last the sheep was released from the glacier's snout after a journey of 1,936 feet.

It was virtually intact and supported eerily in a standing position by a pedestal of ice shielded from the sun by the animal's body. So well preserved that 10 species of plant could be identified in its stomach, the defunct mountain sheep presented the startled naturalists with dramatic confirmation of an old Icelandic adage: The glacier gives back what it takes.

This extraordinary movement of solid ice, a flow that resembles a river in extreme slow motion, gives glaciers their majesty and fascination. It is a stately power; the fastest glaciers known have been gauged at not much more than two and a half miles per year, and some cover less than $\frac{1}{100}$ inch in that same amount of time. But no matter how infinitesimal the flow, movement is what distinguishes a glacier from a mere mass of ice.

For all their great diversity of shapes and sizes, glaciers can be divided into two essential types: valley glaciers, which flow downhill from mountains and are shaped by the constraints of topography, and ice sheets, which flow outward in all directions from domelike centers of accumulated ice to cover vast expanses of terrain. Whatever their type, most glaciers are remnants of great shrouds of ice that covered the earth eons ago. In a few of these glaciers the oldest ice is very ancient indeed: The age of parts of the Antarctic sheet may exceed 500,000 years. But in some small, fast-moving glaciers the oldest ice may have formed no more than a few thousand years ago—though the glacier itself may have existed for 40,000 years or more.

Glaciers are born in rocky wombs—typically a hollow on a mountainside, perhaps even a bowl-shaped cirque scooped out many years before by a previous glacier—above the snow line, where there is sufficient winter snowfall and summer cold for snow to survive the annual melting. At the Poles, the snow line is virtually at sea level. In the Olympic Mountains of the northwestern United States, roughly midway between the North Pole and the Equator, the snow line lies at about 5,500 feet. At the Equator in East Africa it is between 17,000 and 18,000 feet, yet two mountains—Mount Kilimanjaro and Mount Kenya—provide a suitably cool haven for glaciers.

The long gestation of a glacier begins with the accumulation and gradual transformation of snowflakes. Although snowflakes take an almost infinite variety of shapes, each is basically a skeletal six-sided crystal. Soon after it falls to the ground, the delicate extremities of its framework disappear. If the temperature is warm enough, these extremities melt; if the temperature is several degrees below freezing, they evaporate without melting—a process known as sublimation. Either way, complex snowflakes are soon reduced to compact, roughly spherical ice crystals—of the kind referred to as corn snow by skiers. If the crystals survive the melting of the following summer, they are classified by glaciologists as firn—from the German word meaning "of last year."

Naturalist Matthew E. Beatty, of California's Yosemite National Park, inspects the remains of a long-extinct mountain sheep he discovered on Lyell Glacier in 1933. The carcass had been preserved in the ice for perhaps 250 years.

These crystals—usually less than $\frac{1}{10}$ inch in diameter—are the basic components of a glacier. As new layers of snow and firn accumulate, they squeeze out most of the air bubbles trapped within and between the crystals below. The process is signaled by a change in color, from the characteristic white of the airy snow crystals to the steel blue of the compressed crystals of glacier ice. Remaining air spaces may be further reduced by meltwater that trickles down from the warmed surface and refreezes in the frigid depths, often forming larger crystals of ice.

This process of recrystallization into more compact yet larger crystals continues long after the birth of a glacier. In some places crystals of ice may be as big as soccer balls.

The length of time required for the creation of glacier ice depends mainly upon the temperature and the rate of snowfall. In Iceland, where snowfall is heavy and summer temperatures are high enough to produce plenty of meltwater, glacier ice may come into being in a relatively short time—say, 10 years. In parts of Antarctica, where snowfall is scant and the ice remains well below its melting temperature year-round, the process may require hundreds of years.

The ice does not become a glacier until it moves under its own weight, and it cannot move significantly until it reaches a critical thickness—the point at which the weight of the piled-up layers overcomes the internal strength of the ice and the friction between the ice and the ground. This critical thickness is about 60 feet.

Glaciers are often referred to as rivers of ice, but the dynamics of their movement are quite different from those of water. For one thing, glaciers are obviously brittle, as the crevasses that often lace their surfaces clearly testify. There have been many attempts to understand the nature of ice movement in the century and a half since glaciers first came under serious scientific scrutiny. Some researchers thought they were seeing the behavior of a highly viscous fluid, others that the process involved constant fracturing and refreezing. Indeed, it was said that the debates themselves generated enough heat to melt a glacier.

Early measurements of glacial movement used stones or stakes lined up on trees or other reference points alongside the glacier. Others were unplanned: One pioneering Swiss researcher who had built a hut on the ice in 1827 found when he returned three years later that it had moved 120 yards down the glacier. Later methods made use of such sophisticated tools as electronic surveying equipment, time-lapse cinematography and even satellites. But meticulous ground surveys—ever more numerous and accurate with the passing of time—proved to be the key to understanding the complex processes of glacier movement.

Not until the 1950s did researchers agree that there are two basic kinds of processes involved—basal sliding and internal deformation. Both, they found, are present in almost all glaciers, but basal sliding is the predominant mechanism of some glaciers while internal deformation is the main—sometimes the only—way that others move.

Basal sliding, as the name implies, refers to the movement of the glacier over its bed under the pull of gravity. Deceptively simple in concept, the mechanism is not well understood, and has some surprising aspects. For one thing, basal sliding cannot occur without the presence under the glacier of a thin film of water, perhaps only a tiny fraction of an inch thick. The meltwater, provided by heat from the friction of the ice against its bed, or even from the earth itself, serves as a lubricant, reducing the friction of ice against bedrock. However, the extreme cold of the polar regions may prevent meltwater from forming at the bottom of a glacier, so that it remains literally frozen to its bed and can move only by internal deformation.

Another ingredient of basal sliding, known as regelation, enables the ice to bypass small obstructions. When the glacier encounters a pebble in its bed, for example, the pressure of the tiny obstacle's resistance lowers the melting temperature of a minute quantity of upglacier ice until it liquefies. The water flows around the pebble and then, with the pressure removed, freezes again on the pebble's downglacier side and continues to slide along as ice.

The second principal mechanism of glacier movement, internal deformation, is far more complex than basal sliding. Until the 1940s, the prevailing notion was that glacier ice flowed like an extremely stiff liquid. Then researchers began to suspect that glacier ice was behaving much as certain minerals and metals do when under stress at temperatures not far below their melting points. Such materials shatter easily in response to quickly delivered force—that of a hammer blow, for instance—but can at the same time deform and actually flow plastically. During the 1940s and 1950s, laboratory tests established that this process of deformation, often called creep, does occur in glacier ice; the flow of glaciers therefore follows laws of physics that govern the bending of iron or rock, and not the flow of water.

The weight of the ice itself provides the stress that causes glacier ice to creep. Under stupendous pressure, the crystals of ice rearrange themselves in layers of atoms more or less parallel to the surface of the glacier. These layers then begin gliding over one another. The incremental movement of the layers of atoms within each crystal—plus some slippage between crystals, a phenomenon not yet fully understood—constitutes the internal deformation.

The movement of glaciers usually combines the two basic processes in a mix that can vary widely according to local circumstances. Basal sliding can account for almost all of the movement of certain types of temperate-zone glaciers, and for virtually none of some polar glaciers. In 1962, two American investigators, Barclay Kamp and Edward LaChapelle, dug a tunnel under the Blue Glacier on Mount Olympus in the state of Washington. The total velocity of the glacier at its upper surface was only about seven inches a day, and the researchers had to use micrometers to find that fully 90 per cent of its movement was due to basal sliding. But in a similar study done on the Meserve Glacier in Antarctica, where the bottom temperature is −20° F., researchers found absolutely no basal sliding.

The speed of both creep and basal sliding varies markedly according to the thickness, slope and temperature of the ice. The thicker the glacier the faster it deforms, because the increased stress of gravity's pull causes the crystals of ice to creep more rapidly. It has been estimated that an increase in thickness of only a few per cent can increase the speed of glacier deformation by up to 20 per cent.

The slope of a glacier's surface obviously has an enormous effect on its speed. In temperate zones, a steeply inclined valley glacier flows much faster than one on more level ground. The same basic principle applies to polar ice sheets, which advance primarily by internal deformation, even though there is little or no slope to the ground on which they rest. Their velocity is affected by the downward slope of their upper surfaces from the great mound of ice at midglacier. So irresistible is this gravitational force that a polar ice sheet will even cover a mountain in its path so long as the central ice dome is higher than the intervening mountain's peak.

Temperature—the third major factor affecting the speed of glacier movement—varies so greatly that glaciologists sometimes classify glaciers as cold or warm. The warmer the glacier the faster it moves, and the reason is the presence of meltwater. In the warm glaciers of the temperate latitudes, water melted on the surface during summer trickles into the depths of the glacier, refreezes, and gives off minute amounts of latent heat in the process. The heat raises the temperature of subsurface ice, weakening the ice crystals until they yield more readily to stress, and the rate of creep is increased. Another reason for the greater speed of warm glaciers is the presence of greater quantities of meltwater beneath them to lubricate basal sliding.

The difference in the velocities of warm and cold glaciers of comparable size can be dramatic. The five-mile-long Franz Josef Glacier in South Island, New Zealand, creeps and slides at a combined speed of about a yard a day, or a bit more than a fifth of a mile each year. But the ultracold Meserve Glacier of Antarctica, frozen to the bedrock, can only creep, not

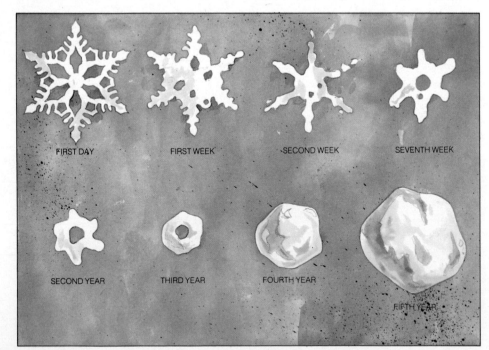

FIRST DAY　　　FIRST WEEK　　　SECOND WEEK　　　SEVENTH WEEK

SECOND YEAR　　　THIRD YEAR　　　FOURTH YEAR

FIFTH YEAR

A series of drawings details the transformation of new-fallen snow into glacier ice. Soon after falling, the snowflakes lose their extremities and are compressed into granules containing air bubbles. The compacted granules, known after one year as firn, are welded together by the pressure of subsequent snowfalls and by the trickling of surface meltwater to form ever larger crystals of glacier ice.

Viewed under polarized light, a cross section of glacier ice from Greenland displays separate ice crystals more than 2,000 years old. The superimposed centimeter grid shows that the large crystal in the center measures about 2.5 by 2.5 centimeters, or one square inch.

slide, and progresses at a rate of scarcely one quarter of an inch per day. It takes three months to travel a yard, and more than 90 years to traverse one fifth of a mile.

Velocity varies not only between glaciers, but within sections of an individual glacier. The first researcher to document this was Louis Agassiz, the 19th Century Swiss naturalist who helped establish the science of glaciology. To study movement in the Unteraar Glacier in the Swiss Alps, Agassiz planted rows of stakes across the breadth of the ice in line with landmarks on the rocky wall of the valley. When Agassiz returned to the glacier the following year, a strange thing had happened. Though all of the stakes had moved a considerable distance downglacier with the flow of the ice, the stakes in the middle of the glacier had moved the farthest. The line of stakes formed a crescent, indicating that the glacier was moving faster in the center than at its sides. Agassiz reflected that water in a stream shows the same tendency to flow more slowly along its banks than in the middle, and he correctly deduced that a glacier did so for the same reason: the drag caused by the friction of the ice against its rocky channels.

With his stakes, Agassiz also established another important characteristic of velocity within a glacier. The ice moves slower at the head of the glacier, where new firn and ice are constantly being formed, and at the terminus than it does in between. Other investigators later found that the fastest flow in a glacier occurs at the so-called equilibrium line. This is the boundary between the accumulation zone upglacier, where more snow falls each year than melts or evaporates, and the ablation zone downglacier, where the annual loss by melting exceeds the snowfall.

There are also variations in the rate of flow at different vertical levels in the ice. Researchers working in the Tyrol, near the border between Austria and Italy, discovered this fact more than 70 years ago by laboriously boring holes through the entire 660-foot thickness of a glacier. They inserted rods in these boreholes and, over the years, found that the rods tilted downhill, indicating that ice in the upper half of the glacier was moving more rapidly than ice at depth. The difference apparently was due to the friction of the underlying rock that slows the ice at the bottom.

Since then, glaciologists have used a variety of drilling methods to probe the movement within many glaciers in all parts of the world. At first, straightforward mechanical bits like those used for drilling oil and water wells were used. To get deeper faster, however, various heat drills were introduced—hot water streams, electrically heated drills and high-velocity gas jets.

As the scientists expected, the flow of ice does not parallel the surface of the glacier. In the accumulation zone, individual ice crystals move at a slight angle downward from the surface as successive layers of firn and ice bury them and compress the underlying levels. In the ablation zone, where the surface of the glacier is losing material to melting, the process is reversed, and the ice crystals angle upward toward the surface. Thus, the oldest ice in any glacier is that exposed near the surface at the snout —where the mummified mountain sheep appeared in Yosemite National al Park in 1933.

The various velocities within the glacier set up a multiplicity of stresses that are for the most part invisible—until they exceed the ability of the ice to withstand them and they suddenly appear as crevasses. These cracks, ranging in size from a fraction of an inch to 100 feet across, occur only in the brittle upper crust of the glacier. At depths of more than 100 feet the ice is under so much pressure that it deforms plastically and does not crack; hence, crevasses rarely exceed that depth. But anywhere the surface velocity

The two zones of a glacier are revealed by the differing visibility of crevasses in this aerial view of Washington's South Cascade Glacier. In the accumulation zone, the crevasses are partially concealed because lower temperatures preserve much of the annual snowfall. But crevasses are clearly evident in the ablation zone, where melting has occurred.

of the glacier changes rapidly across a small distance, crevasses can appear—posing grave hazards to anyone who ventures onto the ice, especially when the chasms are concealed by a fragile layer of snow.

A transverse crevasse—across the direction of the ice's flow—occurs where there is a sudden change in the slope of the underlying bedrock and the glacier speeds up on the steeper slope. The brittle top crust cannot handle the stress, and cracks open. Longitudinal crevasses can form wherever zones of differential movement exist—for example, along the edges of a glacier, which move more slowly than the middle. Thus, some parts of the surface may actually be checkered by crevasses that not only cross the ice but also lie parallel to its direction of flow.

This crisscross pattern is most pronounced—and most perilous for glacier explorers—at the places of accelerated flow known as icefalls. As the name suggests, an icefall somewhat resembles a waterfall. Where the course of a glacier plunges over a steep slope, the velocity of the ice can suddenly triple—to as much as six inches an hour for a distance of a few dozen yards. The brittle surface ice cannot keep pace with the plastic flow underneath and repeatedly splits open under the stress into a honeycomb of intersecting crevasses. As the fissures deepen, they cleave out individual pillars of ice

Both altitude and topography determine whether glaciers form in mountainous terrain. In this diagram, the top of the mountain on the left is below the snow line; hence snow does not accumulate from year to year. The mountain on the right rises above the snow line but is too steep to harbor a glacier. On the center mountain, however, the dual requirements are met, and a glacier descends from the summit.

SNOW LINE

called *séracs*, which eventually can stand more than 70 feet high. These *séracs* are exceedingly unstable; the melting action of sun, wind and rain, the movement of the ice underneath and the pull of gravity often combine to topple the ice towers down the slope in great roaring avalanches of exploding ice fragments.

Icefalls are among the most treacherous parts of a glacier. The fall may appear dormant, but then, in a matter of hours, it can undergo unpredictable and dangerous transformations. In 1951, a party of mountain climbers, including the famed Sir Edmund Hillary, was making a reconnaissance of Nepal's Khumbu Glacier, which later served as the route for Hillary's successful ascents of Mount Everest. The expedition was carefully picking its way across an icefall whose forest of *séracs* had long since been toppled and shattered into a silent ruin. Suddenly, as one of the members wrote later, "there was a prolonged roar and the surface on which we stood began to shudder violently. I thought it was about to collapse." The upheaval ended harmlessly before the climbers could even react sensibly to their peril, and they hastily made their way to safety on solid ice below the fall. Another expedition 18 years later was less fortunate; six of its members died when the Khumbu icefall suddenly roared into action and the brittle surface collapsed.

The speed of a glacier varies not only from place to place but also—as the unpredictability of the Khumbu icefall demonstrated—from time to time. The seasons naturally play an important role. The accumulation area on a glacier may tend to flow faster during winter because snow is piling up, increasing the thickness, and thus the weight and the slope of the ice. By contrast, the area toward the terminus often moves faster in summer because of warmer temperatures, which enhance both the creep of the ice and basal sliding. Seasonal changes in speed of up to 20 per cent are common.

Glaciologists have been able to analyze many of the rhythms of the world's glaciers, but they remain mystified by aspects of other, much faster changes—in particular, the astonishing phenomenon known as glacier surge. When a glacier surges, it seems to defy the findings of a century and a half of research. Instead of creeping and sliding and moving onto new ground at a measured pace, the glacier turns on an extraordinary burst of speed, flowing at 100 or more times its normal velocity. During one of these sudden rampages in 1964, the Muldrow Glacier on Alaska's Mount McKinley was clocked at 48 feet an hour. Such surges are usually short-lived, but between 1936 and 1938 Spitsbergen's Brasvellbreen Glacier set an all-time record when it surged a total of 12 miles.

Glaciologists have only recently begun to study this strange phenomenon—though there is ample evidence of surges in Alpine glaciers just a few centuries ago. The archives of the Chamonix parish in France contain a 1642 account of a glacier rushing ahead for a distance equivalent to "a musket shot" every day. But scientists remained generally unaware of such incidents until the 20th Century and the spread of modern communication into wilderness areas.

The first real attention came in the fall of 1936 when the Black Rapids Glacier in central Alaska suddenly ran amuck. The 12.5-mile-long valley glacier had been retreating slowly. Its terminus was a few miles from the Rapids Roadhouse, a well-known hunting and fishing lodge on the Richardson Highway, at the time the only all-weather highway connecting interior Alaska with the port of Anchorage to the south. Only the caretaker, Mrs. H. E. Revell, and her family were in the lodge that November when rumbling sounds began emanating from the mountains and the lodge began to shake periodically. But Alaskans are accustomed to earthquake activity, and the Revells had no reason to assume that these sounds and tremors were anything different.

Then, on the morning of December 3, 1936, Mrs. Revell was idly scanning the valley above the lodge with binoculars when she was startled to see a huge jumble of ice blocks at the foot of the glacier. The front of the glacier had been relatively placid before; but now it seemed to be in an agony of movement.

During the next days and weeks the Revells watched with increasing apprehension as the glacier advanced implacably toward them. Its terminus was a cliff of ice, 300 feet high and more than a mile wide, from which enormous blocks of ice hurtled down onto the gravelly plain in its path. Within two months the glacier was clearly threatening not only the lodge but the Richardson Highway as well. Around the world, millions of newspaper readers and radio listeners hung on every riveting report from Mrs. Revell. "Black Rapids Glacier sets a speed record," headlined *The New York Times* in a front-page story on February 23, 1937. "Expert finds movement is 220 feet a day."

By March 7, 1937, the tremendous surge of the Black Rapids Glacier —four miles in three months—had brought the ice to within a mile and a half of the roadhouse and the highway. Then, the glacier simply ground to a virtual halt, leaving scientists at a loss to explain the forces that had set it in motion.

Even more spectacular surges have been reported in the Karakoram, the mountainous region of northern India and Pakistan. Early in this century, the Garumbar Glacier is said to have surged so rapidly that it overwhelmed and buried two elderly women attempting to flee it. Although, as a British researcher dryly noted, "one need not credit all the lurid details," a number of fantastic surges have been documented in the Karakoram. One of them occurred in 1953, when the Kutiah Glacier in northern Pakistan sprinted nearly eight miles in less than three months—an average speed of 369 feet a day and the swiftest advance on record.

As in the case of the Kutiah's record-breaking dash, surges typically occur in such remote regions that glacier specialists seldom reach the area until the action is over. But during the 1960s, vastly increased aerial surveillance and photography of glaciers made it possible to observe several major surges, including a modest dash of about 4,000 feet in three years by Alaska's 127-mile-long Bering Glacier—North America's largest.

A worker secures diagonal supports to an accumulation gauge on Switzerland's Aletsch Glacier. In addition to measuring snowfall, the gauge moves with the ice, enabling scientists to measure the glacier's rate of flow.

By the late 1960s, Norwegian engineers had tapped several of their country's glaciers for meltwater to run hydroelectric generators. And the Bondhus Glacier in southwestern Norway seemed an ideal source of additional water for the nearby station at Mauranger. But there was one daunting problem: The station's collection reservoir, Lake Mysevatn, was 1,500 feet above the glacier's snout, from which meltwater was most readily available. To provide enough pressure to drive the generators, it was essential that the water fall all the way from the lake to Mauranger.

The cost of building a pumping station to lift the water to Lake Mysevatn was prohibitive, so the engineers decided to tap the meltwater under the glacier above the lake. They excavated a horizontal tunnel through the bedrock under the glacier, and drilled vertical shafts upward into the ice until they found a natural drainage channel. Then they blasted another tunnel into the rock to channel the glacial runoff to the reservoir. The novel project, completed in

1978, was an immediate success; the Bondhus was soon yielding as much as 15.8 billion gallons of water in a single summer—enough to produce a year's supply of electricity for 6,000 Norwegian households.

A year later, glaciologists decided to take advantage of the inviting tunnel system under the Bondhus to scrutinize the effects of moving ice on bedrock. Approaching the glacier from underneath, they melted caves in the ice and installed sensors to measure the changes in temperature, pressure and bedrock deformation caused by the glacier's movement.

Despite the productivity of the Bondhus hydroelectric project, experts predicted that it might not operate at a profit for many years because of its high maintenance costs. Nevertheless, similar plans to tap other Norwegian glaciers from within were begun, because glaciologists were confident that the knowledge they reaped from their research inside the Bondhus would enable them to drain other glaciers more economically.

A three-story research facility nicknamed The Window perches on a mountain overlooking Norway's Bondhus Glacier. The engineers who harnessed the glacier's meltwater for hydroelectricity used tunnels that extend from The Window to areas underneath the ice.

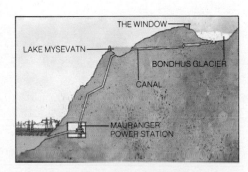

A diagram of the Bondhus project shows how a subterranean canal diverts meltwater from midglacier. The water flows to Lake Mysevatn and then on to Mauranger's generators.

The enormous pressure exerted by the 525-foot-thick Bondhus Glacier passing overhead split a round rock in the wall of a subglacial tunnel. The measuring stick has been placed to show the boundary between ice and bedrock.

A technician uses warm water to melt a cave in the bottom of the glacier. To avoid the danger of a sudden inundation, this work is performed only in winter, when the flow of meltwater through the glacier dwindles away to nothing.

Inside a completed ice cave, a workman installs a structure that resembles rock shaped by glacial pressure. Sensors encased in cement along the frame will record temperature and pressure as the ice reclaims the cave.

Workers inspect an array of sensors that have been cemented directly to the bedrock. The instruments transmit data on the effects of advancing ice to automatic recording devices set up in nearby underground tunnels.

The most closely watched rogue of all was the Steele Glacier in Canada's Yukon Territory. On July 21, 1966, pilot Philip Upton was flying over Mount Steele when he looked down at the glacier on the 16,644-foot mountain's east slope and did a double take. The stream of ice, roughly the size of New York's Manhattan Island, was churned into crevasses and skyscraper-like pinnacles that towered 250 feet into the air. Upton was chief pilot for the Icefield Ranges Research Project, a joint U.S.-Canadian expedition studying glaciers in the region, and he knew that he was witnessing an extraordinary event.

A few days later researchers from the Icefield Project flew in by helicopter. Soon, 30 observers from the United States and Canada were on the scene, launching the first long-term study of a mass of ice on the loose. When the Steele obliged by spurting ahead at two feet per hour, the researchers dubbed it the "galloping glacier."

The galloping glacier had quite an audience. In addition to the scientists, tourists chartered planes to fly over and have a look; bus tours along the nearby Alaska Highway stopped to point out the Yukon's newest natural wonder. And the Steele put on quite a show, bursting over the moraines along its sides, cutting into mountains, shearing off tributary glaciers and damming one river to create a lake. Ice caverns opened up beneath the glacier, some of them large enough to drive a truck through. In a year, the Steele galloped about six miles. Then it simply stopped and resumed its normal behavior, advancing at a modest few yards per year.

From their study of the Steele and many other surging glaciers, scientists put together an increasingly detailed portrait of the phenomenon. Usually, they found, the surging glacier had been fairly stable beforehand, perhaps even retreating. But ice had been accumulating in the upper or middle portion of the glacier. Austin Post of the U.S. Geological Survey, after a study of more than 40 surging glaciers, termed this build-up an "ice reservoir." Over a period of years the reservoir would grow, and for some reason it was not transferred downglacier by the normal mechanisms of flow. When it reached a critical threshold, however, the ice surged down the glacier, setting off high velocities in the existing ice, buckling the surface into characteristic crevasses and jagged pinnacles, thickening the glacier downstream and pushing the snout ahead onto new ground. The surge might continue in full force for as long as three years before it ended and the ice in the lower end of the glacier began to stagnate and melt.

But as the surge ended, another reservoir of ice would start to build, setting the stage for yet another rapid thrust forward. Researchers have found that, in North America at least, most glaciers that have surged have done so periodically—in cycles that range in duration from 15 to 100 years. By studying the clues left behind—principally tangled patterns of morainic loops and churned-up surfaces—Post and other researchers have identified no fewer than 204 glaciers in Alaska and Canada that have surged in the past, and presumably will burst into wild action again sometime in the future.

For all their new knowledge about these unusual glaciers, scientists still cannot predict a surge with any precision. It was long thought that an ice reservoir fills up because of increased accumulation of snow on the glacier—either from several years of unusually high snowfall or from earthquake-induced avalanches. But 15 years of research have failed to establish any consistent link between surges and such external causes as climate or earthquakes.

The most plausible explanation for the extraordinary velocities attained during surging is a vastly speeded-up process of basal sliding. Some re-

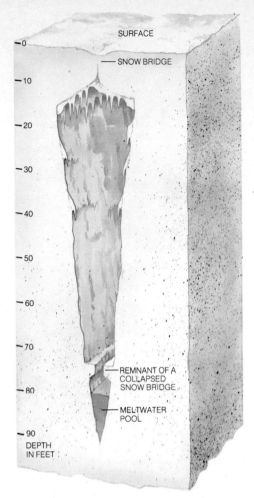

Seen here in cross section, a typical snow-covered glacial crevasse plunges 90 feet beneath the surface and terminates in a pool of meltwater. Snow bridges are created when wind-blown snow forms a cornice on one side and then gradually extends the outcrop until the entire crevasse is covered.

searchers believe that an unusual amount of meltwater forms at the bottom of a glacier, in effect greasing the skids so that the glacier comes uncoupled from its bedrock base. The large amounts of water that flow from under a surging glacier give some credence to the idea. But even if enhanced basal sliding should prove to be responsible for speeding the ice along, other mysteries remain. Chief among these is why some glaciers surge and others in similar settings apparently do not. Another question is why the reservoir of ice builds up in the first place instead of being carried downglacier at a fairly constant rate, as is the case with normal glaciers.

The search for causes is of more than academic interest. Though most surging glaciers are situated in sparsely populated regions where they seldom directly menace civilization, there have been catastrophic side effects. For example, twice in the last three decades, the Medvezhiy Glacier of the Pamir Mountains in the Soviet Union has surged, damming up rivers and then, when the ice barrier melted, setting off destructive floods in an inhabited valley nearby. Surges may similarly be responsible for glacier floods in South America and in the Karakoram Range of Central Asia, where sudden bursts of meltwater are a continuing threat to the new highway linking northern Pakistan and China.

All such effects pale by comparison, however, with the possibility that the biggest glaciers of all might suddenly surge into the oceans. Some glaciologists suspect that large parts of the Antarctic ice sheet surged into the sea thousands of years ago—and may do so again in the near future, triggering a catastrophic rise in the worldwide sea level. It is no wonder that a committee of the U.S. National Academy of Sciences has labeled the enigma of surging glaciers "perhaps the greatest problem to be solved in the field of glacier dynamics."

With the obvious and dramatic exception of surges, the overall result of variations in a glacier's flow is the maintenance of a more or less constant thickness of ice. If the snowfall over a glacier increases, the glacier thickens; the additional weight causes an increase in velocity that carries away the extra material. The process continues all the way down the glacier, and its snout may thus advance a considerable distance while the ice thickness of most of the glacier remains relatively constant. The mechanisms of creep and basal sliding keep a fresh supply of ice moving downglacier to replenish that which is lost through melting and other forms of ablation. If more ice flows downglacier than is needed to maintain equilibrium, the glacier snout advances onto new ground; if less, the glacier retreats. The ice does not actually move backward, of course—the flow is always forward—but the glacier shrinks, its snout surrendering ground it formerly had covered.

To describe the dynamics of growth and recession, glaciologists have borrowed from the terminology of economics such terms as net budget, positive and negative balance, surplus and deficit. Thus, the revenue of the glacier's annual budget is accumulation—the nourishment it receives from falling snow and other sources of moisture. In maritime climates such as that of Iceland or the west coast of upper North America, the mountains catch moisture-laden winds that have passed over a warm ocean current. Here, glaciers may receive more than 30 feet of snowfall every winter. In the interior of a continent, where much less snow falls, glaciers often have to depend in part on snow avalanches from higher slopes or drifts carried in by wind. In central Antarctica, an area of almost a million square miles that has been referred to as a technical desert, yearly snowfall amounts to a little less than two inches, less annual precipitation than occurs in parts of the Sahara. Yet the icecap is maintained because in the cold temperatures that prevail in the polar region there is no melting of surface snow and very little sublimation.

The expenditures of most glacier budgets are made mostly through melting. However, glaciers that end in large bodies of water—chiefly the outlet glaciers that carry ice away from the mammoth ice sheets of Greenland and Antarctica—incur most of their losses in calving enormous icebergs. In temperate areas, glaciers lose some of their ice to evaporation and some of their snow supply to drifting. But melting accounts for 90 per cent or more of the typical glacier's losses. On high mountains, direct solar radiation may account for more than 80 per cent of the heat energy that produces the glacier melting; in others, factors such as the temperature of the surrounding air or the presence of warm rainfall can play a more important role.

When accumulation and ablation balance, the glacier remains in equilibrium. More often, there is either a negative or a positive balance, and the glacier either retreats or advances. Aerial reconnaissance can give researchers a rough idea of a glacier's economy; an advancing glacier usually has a steep or vertical terminus, while a retreating glacier has a gently sloping front. But the balance can change from season to season and from year to year, and in order to determine a glacier's total budget with any precision, researchers must make long-term measurements of both accumulation and ablation.

An additional complication is that the advance or retreat of a glacier usually reflects the weather conditions of previous years. In very small glaciers, the snout may respond to an exceptionally heavy snowfall one winter by inching ahead a greater than normal distance the next winter. Most often, however, because of the large mass of the whole glacier compared with the snowfall of any one year, the response to fluctuations in climate will not become evident for several years. It has been estimated that the Antarctic ice sheet requires at least 10,000 years to reflect distinct changes in climate.

Even similar glaciers in the same neighborhood may not advance or retreat simultaneously in response to the same climatic change. Because of individual eccentricities of altitude, topography and local climate, a glacier may recede while its neighbor is on the march. There are three streams of ice in the Three Congruent Glacier in the St. Elias Mountains in Alaska; one is advancing, another is roughly in equilibrium and the third is retreating *(page 63)*.

Many small glaciers will, of course, appear to advance and retreat in rhythm with the seasons. The snout of the Athabasca Glacier in western Canada's Jasper National Park recedes as much as 100 feet during the summer melting season and recovers much of its lost ground during the following winter. But these are minor, cosmetic changes and do not represent the long-term character of the glacier. Over the past century or so, the Athabasca, for example, has failed to recover all of its losses and has receded nearly a mile, at an average rate of approximately 40 feet per year.

The fluctuation of glaciers like the Athabasca over the centuries tells scientists a great deal about trends in climate. Glaciers, observes one author, "are the great long-term witnesses." They not only reflect but seem to magnify climatic trends, making visible through the comings and goings of ice slight variations in snowfall and temperature fluctuations of only a few tenths of a degree that, over the long term, may constitute a pronounced change in climate.

In nature's unwritten records, a key clue to past climate conditions is found in terminal moraines, the ridges of rocky debris deposited by ancient glaciers. The approximate age of a terminal moraine—and thus the date of the glacier's farthest advance—can be determined by measuring the

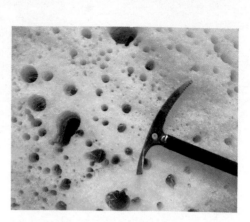

These tubular depressions, called cryoconites, were melted into glacier ice where areas of dark particles absorbed the sun's rays.

As a glacier moved away from its birthplace it wrenched open a chasm, called a bergschrund, in the ice near its rocky headwall.

This medial moraine of rock debris was created near the middle of a glacier when two tributaries joined to form a single ice stream.

Frozen Fantasies on the Glacier's Face

An icefall tumbles down a steep slope in a glacier, where the bending of the ice has broken its surface into giant blocks.

Curved waves of dark and light ice at the base of an icefall, called ogives, evidence the changes in ice density caused by seasonal melting.

A pond formed when summer meltwater filled a depression in the ice absorbs warmth from the sun and will enlarge itself by melting more ice.

Dirt cones appear where sandy deposits on the glacier's surface prevented the ice underneath from melting as fast as the surrounding surface.

The widening of intersecting crevasses when the slope of the glacier's bed increased carved out this towering pinnacle of ice, called a *sérac*.

A torrent of meltwater pours into a moulin, a vertical pipeline in the ice that channels surface water into the body of the glacier.

Being a product of snow, the surface of a glacier might be assumed to present a blank and colorless vista; nothing could be further from the truth. Glaciers have a unique topography, carved by the elements, and heaved and twisted by the stresses in the advancing ice. The results, as the accompanying photographs attest, can be as resplendent as any scenery on earth.

The movement of a glacier accounts for its most awesome features. The brittle surface ice, carried along by the more pliant ice that is flowing, bending and twisting beneath it, often splits open in a multitude of crevasses. At a steep incline, intersecting crevasses can cleave the ice into enormous blocks called *séracs (bottom row, center)*, which may eventually tumble and crash in an icefall *(top row, center)*.

The summer sun is a somewhat more subtle agent of change, but it erodes and reshapes a glacier as implacably as the glacier sculpts the land. As the sun melts the glacier's surface, meltwater gathers in glistening blue ponds *(center)*, or trickles from place to place, cutting channels in the ice. Sometimes the water cascades into fissures in the ice and refreezes, creating a rootlike lacery of light-colored ice in the darker, compressed ice of the glacier's interior.

Occasional dark scatterings of sand or dust on the ice absorb a disproportionate share of the sun's heat and melt their way into the glacier in a shotgun-blast pattern of pockmarks *(top left)*. But if the deposits are thick enough, they insulate and protect the ice beneath them, remaining as dirt-covered humps and ridges when the surrounding ice melts.

As a glacier scours the sides and bottom of a valley, it plucks out and carries along its edges a stream of rock debris called moraine; when two glaciers merge, their moraines can form undulating ribbons of contrasting shades *(bottom left)*.

The surface of a glacier fairly bursts with hue and pattern; the gemlike blue of crystal ice, the gray of ice loaded with rock dust, occasionally even a red blush caused by algae that live on the ice—all contribute to an ever-changing icescape of infinite beauty.

Marbled by contorted medial moraines, the Susitna Glacier in central Alaska offers dramatic evidence of glacial surge—sudden and rapid advance. The illustrations at left, based on before-and-after photographs, show that ice from the lower tributary was displacing the larger flow. Then a powerful surge in the main stream pushed the intruding ice flow downglacier. The 1970 photograph above shows subsequent movement.

amount of radioactive carbon remaining in the fossils of organic material found among the rocky debris. Even if fossils are not found, the measurement of the largest lichens growing on the moraine can yield a rough estimate of the time of deglaciation—up to 4,000 years in the past. Important clues about fairly recent changes can be found in forest trimlines—the boundaries between old trees untouched by the ice and newer ones that sprouted after the glacier terminus receded. These trimlines can be dated by counting the number of annual growth rings in the oldest of the newer trees.

Using the data gleaned by such methods, glaciologists have made extensive studies of the fluctuations of glaciers during the 8,000 years or so since the disappearance of the huge ice sheets that once covered much of North America and Europe. They have found evidence of at least three distinct periods of glacier expansion since the end of the Great Ice Age. The earliest expansion followed a period of worldwide warming, during which temperatures in the middle latitudes averaged some 2° F. higher than present-day temperatures and melted the remnants of the Ice Age sheets in North America and Europe. The subsequent expansion reached its peak about 5,000 years ago, but hardly represented a new ice age: Indeed, the glaciation was only slightly greater than modern levels.

The second period of glacier expansion, which peaked about 2,500 years ago, saw temperatures cold enough to nurture the birth and growth of many new glaciers. In fact, contrary to popular belief, many of today's glaciers are shrunken remnants not of the Great Ice Age but of this second subsequent worldwide cooling.

The last period of expansion began only about 600 years ago and continued unabated for about 450 years. After centuries of glacier recession, many Europeans felt safe in building their homes near the edges of the ice. Then, in the early 15th Century, the ice began a sudden, inexorable advance. At the peak of this prolonged cold spell, glaciers bulldozed farmers' fields in Norway and Iceland, and even invaded some Swiss villages, crushing homes while their desperate occupants vainly chopped and hauled away the encroaching ice. The cold was so severe and the accompanying glacier expansion so pronounced that the period is sometimes called the Little Ice Age.

In about 1850, the most recent advance of the glaciers gave way to a century-long retreat that coincided with a worldwide increase of about 1° F. in mean temperature. In the southern Himalayas, several large glaciers receded two and a half miles. In Montana's Glacier National Park, Sperry Glacier, once the largest of the park's 50-odd streams of ice, shrank to less than a third of its former area.

Nowhere was this recession more obvious than in the European Alps, where by the middle of the 20th Century at least 80 per cent of all glaciers were shrinking. One of Switzerland's best-known glaciers, the Rhone—whose meltwater nourishes the river by that name—has been carefully observed and often depicted by artists during the past three centuries. Old prints dating from the 18th Century show the Rhone proudly advancing into the Gletsch valley. It reached its maximum size about 1818, extending to within yards of a hotel whose guests could wander out the door and onto the ice. Less than a century later, it had retreated one and a half miles back up the mountain slope whence it came.

In the late 1940s, the Northern Hemisphere started getting cooler again, a trend that was promptly signaled by a glacier—the Nisqually on Mount Rainier in the state of Washington. The fluctuations of the Nisqually have

Although they flow into nearly identical, parallel valleys, three Alaskan glaciers show different stages of balance. The glacier on the left has recently advanced, its blunt terminus outdistancing the debris on the valley floor. High lateral moraines flanking the middle glacier suggest that it is receding and has been doing so for a number of years. The third glacier appears to have made a very recent advance, from which it has not yet begun to retreat.

been periodically recorded since 1857, and when the U.S. Geological Survey began systematic photo surveillance in 1942, the researchers found that the glacier had thinned out and lost about 15 per cent of its five-mile length in 85 years. But routine measurements taken in 1946 showed that the Nisqually had suddenly thickened in its accumulation area. This thickening subsequently coursed down the glacier, increasing its velocity by a factor of five to an annual rate of 80 to 100 feet.

Before long, glaciologists in North America, Europe and Asia noted a widespread slowdown in glacier retreat, and numerous examples of expansion. At least a third of the glaciers in Switzerland, including the Rhone, began to inch forward. In Alaska, 7 per cent of 200 glaciers measured were found to be on the march, and while 63 per cent were retreating and 30 per cent were holding their own, the advancing glaciers contained more ice than all the others put together. Although some authors have used this evidence to support predictions of an imminent return of the Ice Age, most scientists agree only that the present interglacial period will eventually end—probably no sooner than 2,500 years in the future.

For all the fears it promoted of another Ice Age, news of the advance of the glaciers was welcome to the millions of people around the world who use glacier meltwater to power their generators and irrigate their fields. In addition, many of the world's great rivers—the Columbia of North America, the Rhine and Danube of Europe, the Indus and Ganges of Asia, even the tropical Nile of Africa and the Amazon of South America—are nurtured by melting glaciers.

After its load of soil and debris settles out, meltwater is exceptionally

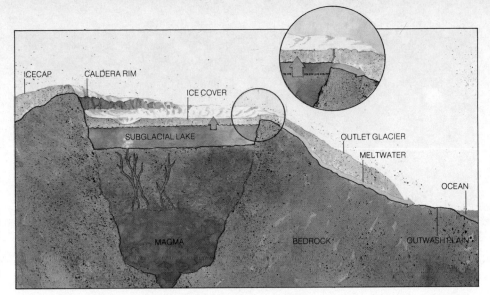

ICECAP   CALDERA RIM   ICE COVER   OUTLET GLACIER   MELTWATER   OCEAN

SUBGLACIAL LAKE

MAGMA   BEDROCK   OUTWASH PLAIN

A cross-sectional diagram of the Grimsvotn caldera shows how a lake of meltwater gradually rises until the overlying ice floats and the lake surges through a low point in the caldera rim (*inset*). After coursing 30 miles through subglacial tunnels, the water emerges from the ice to flood the lowlands.

pure, and it flows through home taps in many cities. Boulder, Colorado, has its own glacier, the Arapaho, which, despite an area of only 100 acres, produces about 260 million gallons of drinking water each year. From the glacier's perch 12,000 feet up the eastern slope of the Continental Divide, 30 miles west of the city, meltwater flows into a chain of mountain lakes from which it is piped to the city. There, residents can drink cold, crystal-clear water from ice that was formed long before the discovery of America.

In some areas of the world, summer glacier melting pours out water in such torrents that farms, highways and human life are in peril. Sometimes a glacier will dam a stream and create a lake; then, during the summer, rain and meltwater may raise the level of the lake until water flows over the ice dam, melts it away and unleashes a flood. Floods of this type occurred every year at Alaska's Lake George, which was alternately dammed by the Knik Glacier and drained away during the summer, until 1966. The process stopped that year when the glacier receded, but it could resume at any time.

By far the most spectacular of these floods are the *jokulhlaups,* or glacier bursts, that spew forth from the icecaps of Iceland. *Jokulhlaups* result when meltwater dammed within or beneath the glacier bursts free. And in the Land of Fire and Ice, they often are associated with volcanic activity.

Two icecaps in Iceland—the Myrdalsjokull and the Vatnajokull—sit atop active volcanoes whose heat is constantly melting the lower parts of the glaciers. In the summer of 1918, so much meltwater built up under the Myrdalsjokull icecap that when it broke open, the torrential flow for a time was triple that at the mouth of the mighty Amazon River.

Since then the Myrdalsjokull has been quiet, but not its larger neighbor, the Vatnajokull. Beneath this 3,280-square-mile icecap, Iceland's biggest, lies a volcanic crater whose heat creates a reservoir of meltwater that covers an estimated 14 square miles to a depth of more than 1,500 feet. Once every five years or so, the water apparently lifts part of the icecap, escapes from the crater and rushes 30 miles under the ice to burst forth onto a sandy, largely uninhabited plain, where it deposits huge boulders and blocks of ice bigger than three-story houses. Such a flood is a truly awesome event; it can turn 386 square miles of the plain into a vast lake and is often accompanied by a volcanic eruption and a stench of sulfur, from the volcanic hot springs, that can kill flocks of birds and wilt acres of foliage.

Only rarely, however, do glaciers release their meltwater with such devastating force. Water usually gurgles from a glacier in a far more benign fashion that is highly inviting to an increasingly thirsty world. An estimated three quarters of the world's fresh water—the equivalent of 60 years of

A 1972 meltwater flood from the Grimsvotn caldera created a braided tracery that extended across 300 square miles of Icelandic lowlands. The caldera unleashes such floods approximately every five years.

rain and snow over the entire earth—is locked in glacier ice. Only a small portion of that ice represents a potential water resource (about 85 per cent of it occurs in remote Antarctica and Greenland, where relatively little melting takes place), but the remaining glaciers hold as much water as all of the world's rivers and lakes. Indeed, the summer meltwater from just the 800 small glaciers of Washington State yields 470 billion gallons of water.

Glaciers provide a convenient system of storage for their water, holding it in the form of ice during winter when the need generally is least. Unlike rivers, which reach peak flow during the rainy spring—glaciers release most water during the hot, dry summer when it is needed most. In addition, glaciers tend to be more stable than rain-fed rivers and require neither expensive dams nor maintenance of any sort. As a reservoir for water, writes one expert, "a glacier is money in the bank, a trust fund that pays reliably, steadily year after year."

In recent years, scientists have started manipulating glaciers in hopes of increasing the payoff from this natural trust fund. In the Soviet Union, researchers have experimented with the idea of blowing up glaciers in order to increase the runoff of meltwater. But most plans concentrate on changing the light-reflecting properties of glaciers. Ordinarily the glacier's white surface reflects much of the radiation reaching it from the sun. Glaciologists use the term "albedo" to describe the amount of solar radiation reflected; fresh, clean snow cover has an albedo of more than 90 per cent. If a dark substance is spread over the surface, the albedo is drastically reduced; the ice will reflect less radiation, absorb more and, thus warmed, will melt faster.

This simple principle—that white reflects solar radiation and colors absorb it—was demonstrated in 1761 by that inveterate American tinkerer, Benjamin Franklin. As an experiment, Franklin placed several cloth squares of different colors on snow in full sunlight. He observed that the black square melted into the snow to the greatest depth, and concluded that it had absorbed more of the sun's heat. The lighter the color of the cloth, the less snow it melted; the white cloth sank scarcely at all. Franklin then used the results of his experiment to advise a woman friend that summer hats ought to be white because they would not absorb the sun's heat.

But Franklin was just discovering what Asian farmers had known for 2,000 years. In the spring, these farmers customarily spread dirt and cinders on the snow to melt it from their fields. And centuries later, peasants in the arid interior of China began applying the same method to glaciers to increase the summer runoff of precious water.

In modern times, dusting of glaciers on a much larger scale has been contemplated by at least four national governments. Experiments aimed at stepping up the flow of glacier-fed streams have been conducted by the United States in the state of Washington and by Chile in the Andes. Soviet and Chinese tests have centered on the complex of Central Asian mountain ranges that includes the Pamirs, the Tien Shan and the Karakoram. Runoff from masses of ice there provides irrigation for arid land not only in China and the Soviet Union but also in the bordering countries of India, Pakistan and Afghanistan. All of these nations desperately need more water to increase food production.

The Soviet researchers who dusted glaciers in the Tien Shan found that coal dust was the most effective darkening agent. The black dust had to be applied in a thin layer, however, because a covering only three eighths of an inch thick was sufficient to insulate the ice, actually slowing down the melting. By spreading coal dust at a rate of up to 25 tons per square mile—by hand and from airplanes—the Soviets were able to increase runoff by as much as 55 per cent. But the cost of transporting the enormous quantities of blackening material remained far too high to permit widespread glacier dusting.

Scientists in several countries have experimented with a different approach to managing a glacier's meltwater runoff: building up the accumulation of snow that will turn to firn and eventually ice. They seeded clouds to increase snowfall, set off explosives to trigger avalanches of snow from higher slopes and erected drift fences to direct blowing snow onto the glacier. Only the snow fences yielded enough additional meltwater to justify their cost.

An impressive estimate of the value of glaciers to modern man can be found in Japan, which has no glaciers even though its mountains harbor more than 200 patches of perennial snow and ice. These mountain basins often build up thick accumulations of firn, but never enough to create a glacier. At one such firn field in the Tateyama Mountains in central Japan, scientists have spent more than a decade studying ways to enhance snow accumulation through the use of drift fences and artificial avalanches and to retard summer melting by covering the area with plastic insulation. Their long-range objective is nothing less than the artificial birth of a glacier.

Not long ago, Professor Keiji Higuchi of Japan's Nagoya University declared that success for the project was still at least a decade away. But he expressed hope that one day Japan would have a number of artificial glaciers; they would provide water to ease the country's frequent droughts, stimulate the tourist industry and even act as giant air conditioners that would moderate Japan's searing heat waves. The advent of Professor Higuchi's artificial ice age is, by his own assessment, problematic: "Our present status is that of conducting the basic experiment on it." Ω

CONTRASTING WORLDS OF MOVING ICE

The very word "glacier" connotes an eternally cold landscape, remote and unchanging. Yet as the map at right indicates, the world's glaciers are not restricted to the frigid polar reaches. Moreover, they are anything but still.

Glacial ice is the product of a decades-long process that compresses airy flakes of snow into extremely dense, steel-blue crystals. Glaciers can be found anywhere that snow accumulates year after year for the required length of time; the controlling factor is not how much snow falls, but how much of it remains after seasonal melting.

Fully 98 per cent of existing glacial ice is found in the polar regions, where the sun's rays strike the earth obliquely and seldom if ever raise the temperature above freezing. Although surprisingly little snow may fall near the Poles (much of Antarctica receives no more than 12 inches per year), virtually all of it survives to add to glacial ice.

The remaining 2 per cent of the earth's glacial ice is distributed among 200,000 glaciers scattered throughout the rest of the world. In temperate latitudes, ice can thrive only on mountains, at higher, thus cooler, elevations. If the mountains are near the sea and receive plenty of precipitation—as is the case with the coastal mountains of southern Norway, for example, where 150 inches of snow fall each winter—glaciers are found as low as 6,000 feet. But the drier American Rockies can support glaciers only at levels closer to 10,000 feet. And along the Equator, glacier-bearing mountains must be even higher to provide a suitably cold environment. Kilimanjaro in East Africa harbors a small glacier—but only above 17,000 feet.

Although the giant polar ice sheets and the far-flung valley glaciers are all formed from snow, they differ in almost every other respect, as the following pages illustrate. Whereas ice sheets totally dominate their geography and even modify the climate by reflecting solar rays back into space, valley glaciers are confined by the surrounding terrain and controlled by the local climate.

A map of the world's glaciers shows the preponderance of the Greenland and Antarctic ice sheets and the wide dispersal of the comparatively tiny valley glaciers. In all, glacial ice covers 5.8 million square miles—or just over 1/10 of the world's land surface.

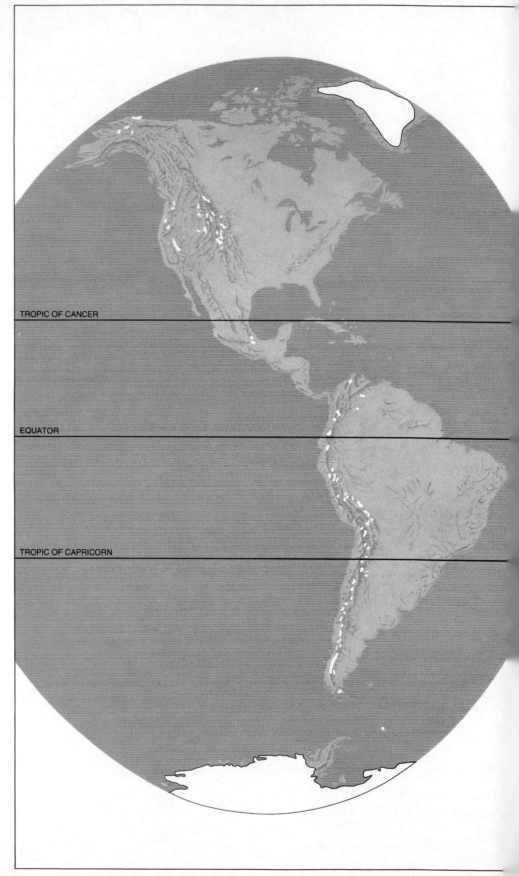

TROPIC OF CANCER

EQUATOR

TROPIC OF CAPRICORN

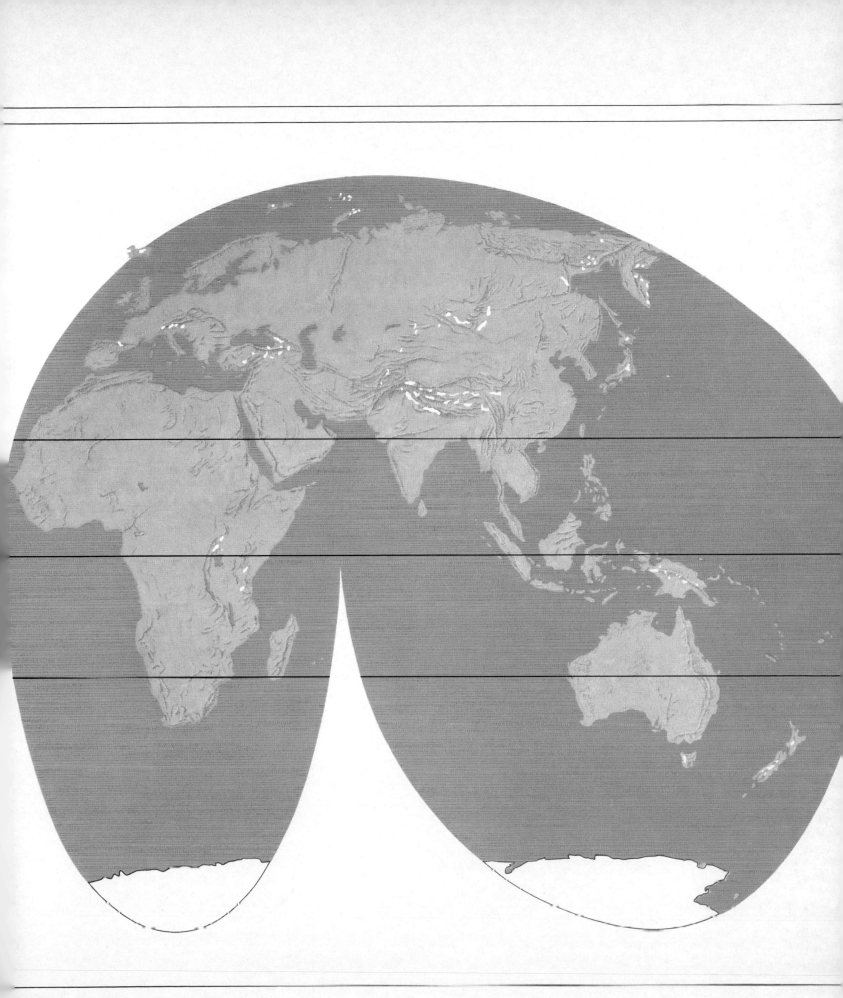

Profile of a Mountain Glacier

Like honey running down a tilted slice of bread, a valley glacier responds to the pull of gravity with two distinct sorts of movement, one operating along its base and the other throughout its bulk. The skidding of its bottom over a slippery film of meltwater that lubricates the underlying bedrock is called basal sliding; internal deformation, or creep, occurs when the glacier's ice crystals change their shape under pressure and slide over one another.

The relative contribution each mechanism makes to the total amount of the glacier's movement varies with the steepness of the slope and the temperature and thickness of the ice. In winter, patches of the upper reaches of a valley glacier may freeze to the bedrock, preventing basal sliding. But in summer, when the ice throughout the glacier is near its melting point, as much as 90 per cent of the total movement may be basal sliding. A sliding glacier has a bulldozer effect and hence enormous erosive force.

The speed of a valley glacier can vary from a few inches to a few yards a day and is greatest where the glacier is thickest. This segment of the glacier is called the equilibrium line, because it divides the accumulation zone of annual snow and ice build-up from the ablation zone, where some of the glacier is lost every year to melting.

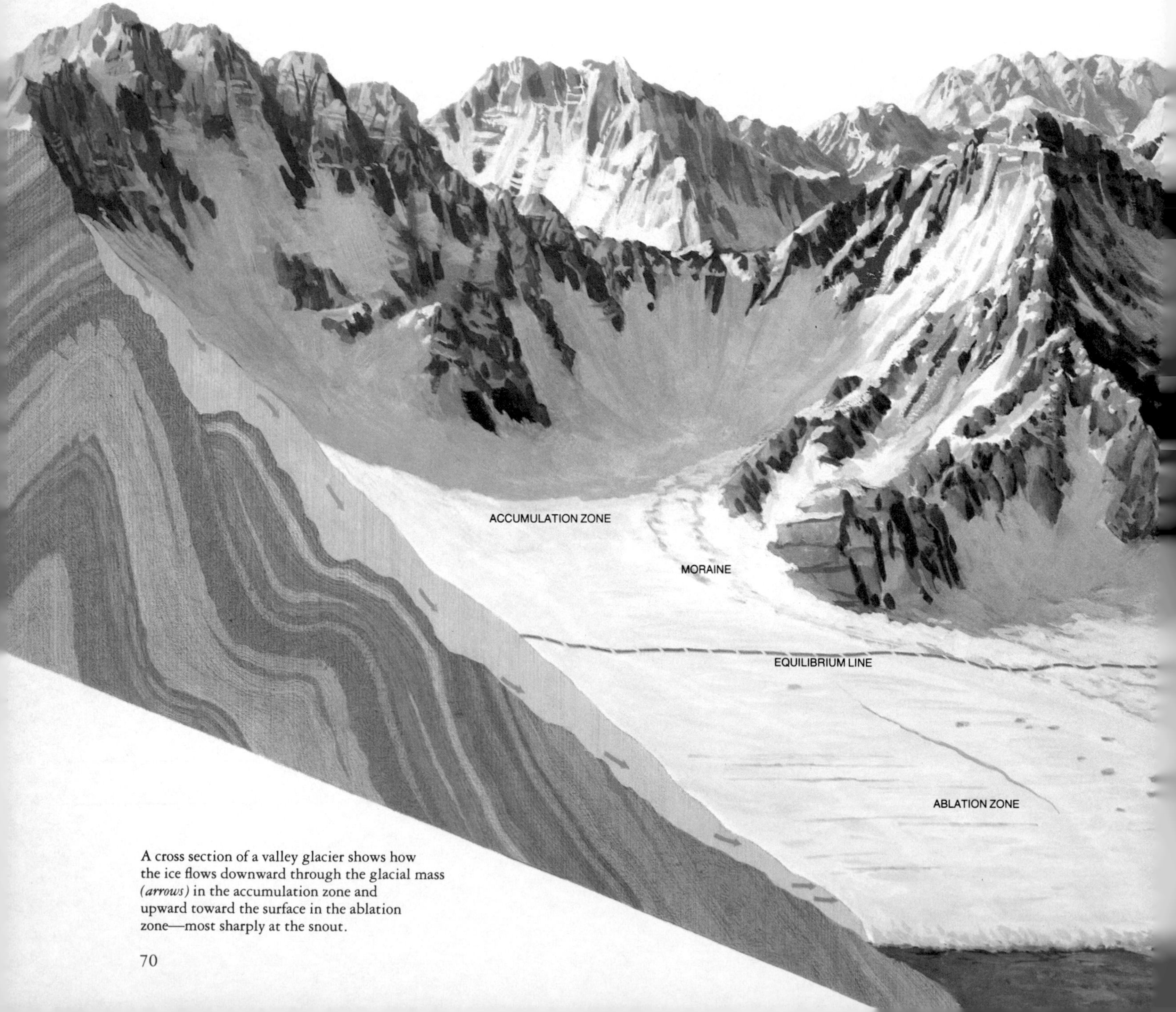

ACCUMULATION ZONE

MORAINE

EQUILIBRIUM LINE

ABLATION ZONE

A cross section of a valley glacier shows how the ice flows downward through the glacial mass (*arrows*) in the accumulation zone and upward toward the surface in the ablation zone—most sharply at the snout.

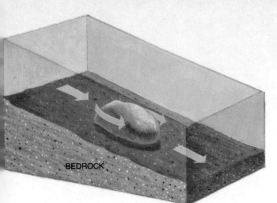

BEDROCK

Glaciers bypass small obstructions in their beds by a process called regelation *(left)*. Increased pressure on the upslope side of the obstruction allows the compacted ice *(blue arrow)* to melt. The water *(green arrows)* flows around the obstruction and then refreezes.

As ice moves over a hump in its bedrock floor, bending stresses open crevasses in the brittle top layers of the glacier. The maximum depth of such crevasses is about 100 feet; below that, the pressure exerted by the glacier's weight keeps the ice flexible.

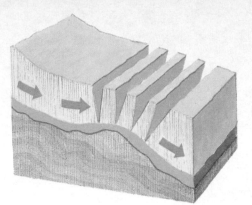

MORAINE

ACCUMULATION ZONE

The Shrouds That Spread from the Poles

The continental ice sheets of Greenland and Antarctica radiate outward in all directions from huge central ice domes that have been built up by snowfall over millions of years. Except for a few mighty mountain peaks, called nunataks, that poke up through the enveloping blanket of white, the ice sheets totally bury the landscape.

Although scientists have found meltwater by drilling boreholes to great depths, so much of the ice sheet is frozen to bedrock that basal sliding is impossible. Instead, as in cold patches in valley glaciers, movement proceeds almost entirely by internal deformation. The weight of the ice generates so much pressure—300 tons per square foot at depths of 10,000 feet—that the crystals realign in the same plane, much like a deck of cards, and begin to slide over one another in minute increments that cumulatively produce a forward flow of the entire ice mass. Velocity is about 30 feet per year.

Unlike valley glaciers, the frigid polar sheets lose almost no ice to melting or evaporation; ablation is achieved almost entirely by the calving of icebergs into the sea. As the ice creeps ponderously off the bedrock into the water, a weak spot eventually gives way and, with a roar, a massive chunk of ice breaks off and floats away to melt in warmer waters.

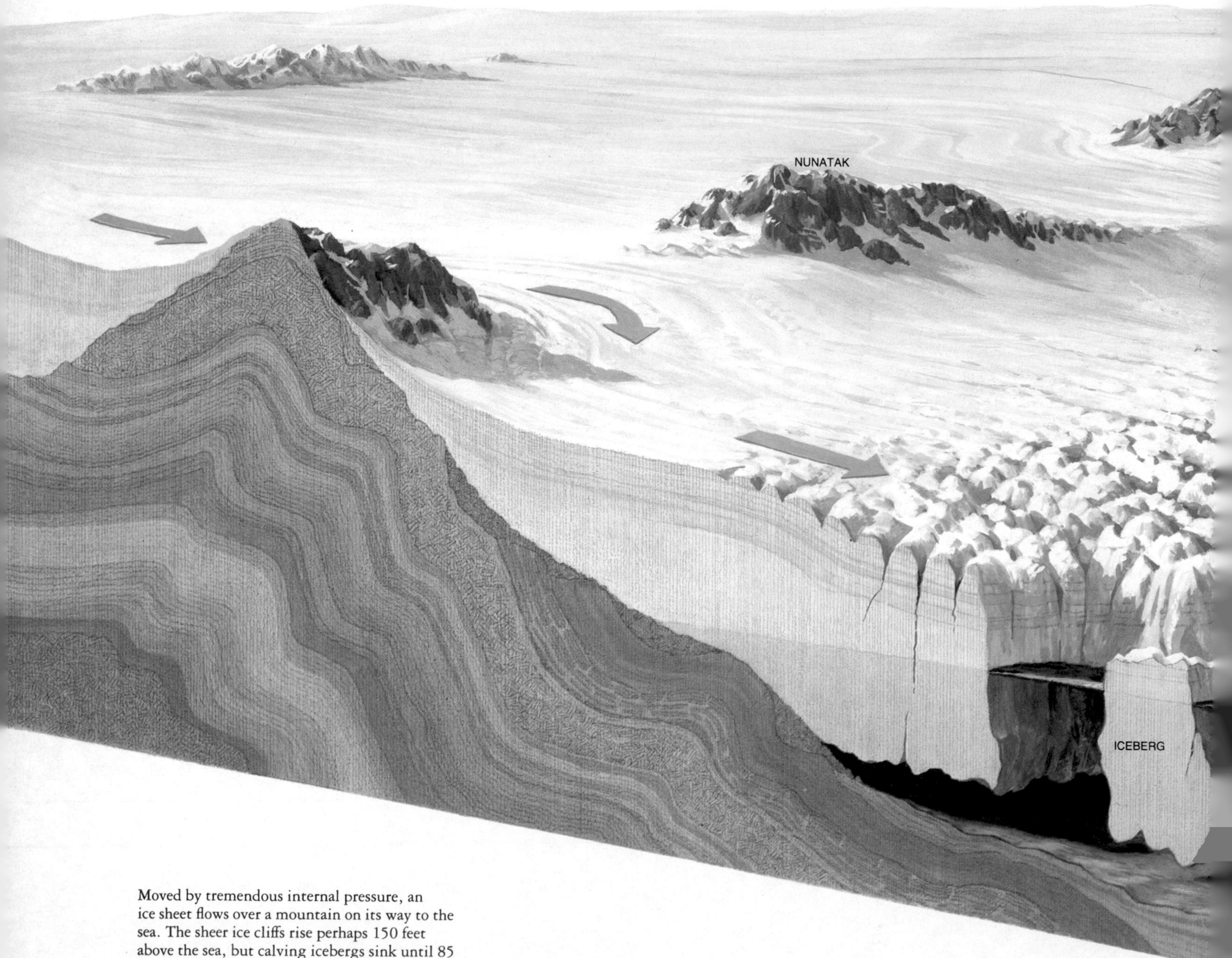

NUNATAK

ICEBERG

Moved by tremendous internal pressure, an ice sheet flows over a mountain on its way to the sea. The sheer ice cliffs rise perhaps 150 feet above the sea, but calving icebergs sink until 85 per cent of their mass is below water.

72

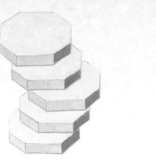

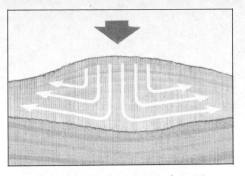

A three-step diagram illustrates the behavior of ice crystals under enough pressure to cause internal deformation. Increasing pressure on a jumble of ice crystals (*left*) causes them to align (*center*) and eventually slide over one another.

The downward and outward path of glacial ice is diagramed in this cross section of the 14,250-foot-thick central Antarctic ice dome. The weight of the dome depresses the earth's crust by as much as 5,000 feet.

ICE CLIFF

"Among the agents that nature has employed in making these mountains there is one that above all others deserves the name of Destroyer, it is the glacier."

So wrote American naturalist John Muir after exploring the glaciers of the Sierra Nevada in California in the latter part of the 19th Century. At the time, the scientific community was still struggling to understand how glaciers moved, let alone how they could carve away mountains. But their sculpting efficiency was indubitable; Muir marveled at ice-grooved canyons, smooth and polished knobs of granite, and angular glacier-carved peaks.

Since Muir's day, scientific scrutiny has confirmed that glaciers are among the earth's mightiest engines of change. During the last two million years, glaciers have enveloped nearly one third of the land area of the planet; their imprints are everywhere; the spires of the Matterhorn, Mount Everest and hundreds of other mountains were fashioned by glaciers, as were the Great Lakes of North America.

The thicker a glacier's ice and the faster it moves, the more effective it is as a natural polishing, cutting and grinding tool. By studying relic landforms, scientists have estimated that the North American ice sheet, which 20,000 years ago covered all of Canada and the United States east of the Rocky Mountains and as far south as Long Island, gouged an average of 30 feet of rock from the surface of more than two million square miles of territory.

Yet for all their destructiveness, glaciers have created many of the planet's most spectacular landscapes, as the photographs here and on the following pages attest. And that was what most deeply impressed John Muir, who said, "Nothing that I can write can possibly exaggerate the grandeur and beauty of their work."

The mountains of Montana's Glacier National Park testify to the sculpting effect of glaciers with their proliferation of cirques, which are deep basins with precipitous headwalls that have been carved by moving ice.

A climber surveys the wintry splendor that surrounds Blea Water, a small lake formed in a cirque basin that was scooped out in England's Lake District by an ancient glacier.

A row of sharp peaks, called horns, soar skyward in the Chigmit Mountains of Alaska. The granite needles were left behind by the carving away of intersecting glacial cirques.

Crescent-shaped gouges left by a glacier scar the bedrock of Bear Mountain in southern New York State. These "chatter marks" may have been caused by the rebounding of boulders that were imprisoned in the moving ice.

A tracery of grooves, cut into bedrock by stones embedded in a glacier, marks the primeval path of the ice across California's Yosemite National Park. The boulders are erratics, transported to this spot by the glacier and left behind when the glacier retreated.

These small hills, or kames, were deposited by
the outwash of a glacier in the Canadian Rockies
in British Columbia. Deposited in separate
layers, with finer materials near the top, kames
often provide a commercial source of gravel.

Rocky debris known as till, left behind by a
melting glacier, blankets a field in Northrop,
Colorado. The limits of such till deposits
delineate the extent of past glaciation.

A glacier left this line of debris, called a lateral moraine, in Euseigne, Switzerland, several thousand years ago. Wind and rain carved away the material not protected by overlying boulders, leaving these fantastical shapes.

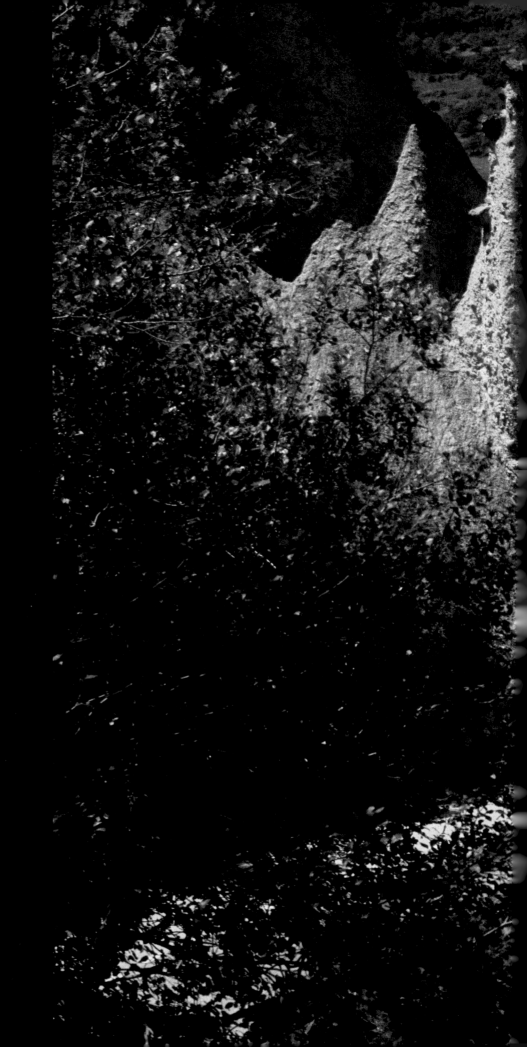

Chapter 3

PHANTOM FLEETS OF THE POLAR SEAS

Describing his first view of Antarctic waters in 1928, Admiral Richard E. Byrd spoke of "stricken fleets of ice bigger by far than all the navies in the world, wandering hopelessly through a smoking gloom." When 19th Century naturalist John Muir encountered icebergs in Alaska in the 1880s, he called them things "of ineffable beauty, in which the purest tones of light pulse and shimmer, lovely and untainted as anything on earth or in the sky." To surgeon Isaac R. Hayes, who watched them by sunset off Greenland in 1860, they seemed "masses of solid flame or burnished metal." To the more prosaic eye of seaman Gerrit de Veer, who accompanied the Dutch navigator Willem Barents on a search for a passage across the top of Russia to the Orient in 1596, icebergs were "great heaps, as big as the salt hills that are in Spain."

As these diverse observations by polar pioneers attest, icebergs are infinite in their variety and widespread in their occurrence. Majestic, mysterious and ever-changing, they can be flat as prairies, craggy as mountaintops, columnar as temple ruins. They have been perceived as black, crimson, gold and purple; as emerald green and cobalt blue; as battleship gray and opalescent white. They girdle the continent of Antarctica, gyre in the Arctic Ocean, lurk off the shores of Greenland, tumble through the fjords of Alaska, intrude on the North Atlantic. An occasional sport among them sometimes strays deep into temperate waters, south from Labrador as far as the latitudes of Florida and the Azores, north from Antarctica to the waters off South Africa.

In the parts of those regions unused by man, icebergs are merely natural curiosities and objects of scientific inquiry; they are studied to determine, for instance, their role in the hydrologic cycle, the continuous exchange of water between the earth's surface and the atmosphere. But as human civilization presses relentlessly into every quarter of the earth, icebergs represent both menace and promise. Since at least the 10th Century, when the Vikings ventured west from Greenland, icebergs have demolished ships and taken the lives of mariners, and in more recent times they have bedeviled seekers after oil—while simultaneously holding out to parched nations the tantalizing prospect of enormous water resources awaiting only new technologies for transport.

In simplest terms, icebergs are blocks of fresh-water ice adrift at sea and floating seven eighths submerged. They originate in the ice shelves that skirt the shores of Antarctica and parts of the Canadian archipelago, and in the glaciers of such regions as Greenland, Alaska, Spitsbergen and Novaya

An iceberg looms out of the mist off the port bow of a fishing vessel in Scoresby Sound, Greenland. Arctic glaciers calve tens of thousands of huge bergs into the polar seas every year, and hundreds of them drift far enough south to threaten shipping and offshore oil rigs.

Zemlya. They come into existence by a process called calving—splitting off from the parent ice. Sometimes the calving results from stresses brought to bear on the ice as it undulates with currents and tides after extending itself into the sea; sometimes it occurs because subglacial runoff has undermined the glacial front and left it without support. In some instances the split takes place along a fissure that formed when the glacier was working its way downslope, over bumps and into depressions and around bends; in others it occurs along the 10- and 12-foot indentations worn by waves constantly thrashing at the water line.

Newborn icebergs range in size from a modest 20 feet in height and 50 feet in length to a height of 250 feet and length of a mile or two; an occasional giant may tower more than 500 feet high and be several miles long. They may live for only a single season or for a decade or more, but sooner or later they waste away. They disintegrate partly by melting but even more by repeating the calving process, producing smaller and smaller progeny known to glaciologists as bergy bits, growlers and brash.

"A berg suddenly going to pieces is a grand sight," wrote John Muir, "especially when the water is calm and no motion is visible save perchance the slow drift of the tide current. The prolonged roar of its fall comes with startling effect, and heavy swells are raised that haste away in every direction to tell what has taken place, and tens of thousands of its neighbors rock and swash in sympathy, repeating the news over and over again."

As icebergs break down they change shape, and for all the myriad forms that poetic-minded observers have seen in their flights of fancy, glaciologists have settled on five broad categories: tabular, blocky, drydock, pinnacled and domed.

Tabular icebergs, as the name implies, are flat-topped, but not necessarily smooth; the surface may be jagged with spurs and rent with crevasses formed when the glacier was still landbound. The distinguishing criterion is not so much their flatness, however, but their proportions. Tabular icebergs are much longer than they are high, by a ratio of at least 5 to 1. Most

Clouds of spray and ice at the base of the 200-foot-high Hubbard Glacier mark the plunge of a newly calved iceberg into Alaska's Disenchantment Bay. An iceberg is launched when the glacier's snout, eroded from beneath by waves and from above by meltwater streams as it advances into the water, eventually fractures.

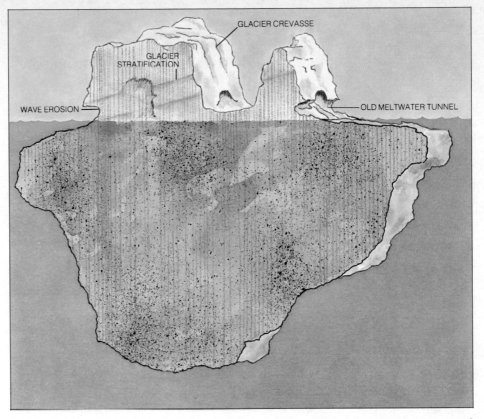

of the bergs that fall into that classification are born of the Antarctic and Canadian ice shelves, but not all; mile-long icebergs are common in the Greenland fjords, and the Columbia Glacier in Alaska sometimes calves tabular icebergs that are 200 feet long. Blocky icebergs, too, are more or less flat-topped. However, their length is no more than two and a half times their height. Such icebergs generally calve from the relatively narrow snout of a glacier protruding into a fjord. But blocky bergs can also result from the breakup of a large tabular iceberg.

Repeated calvings, combined with wind and wave erosion, cause all icebergs to become more and more irregular in shape—begetting the three other categories. Drydock icebergs have one or more U-shaped troughs, formed where wind and waves have bored through a fissure in the iceberg's upper surface. Pinnacled icebergs are generally pyramidal in shape, with fantastic soaring spires that make the icebergs extremely unstable. As their center of gravity shifts, they spin, pitch, yaw and sometimes somersault; the exposure of a new surface or a new pinnacle subjects the iceberg to further turning and further erosion. A domed iceberg, on the other hand, is apt to be the product of a pinnacled or a blocky iceberg that has flipped over and exposed a face made round by rocking in the water.

How and when icebergs calve depend on traits bestowed on them by their parent glaciers. The Amery Ice Shelf in Antarctica calved only once between 1950 and 1980; when it did (in 1963) it delivered itself of 218 cubic miles of ice. Yet some of the Greenland glaciers release about a third of a cubic mile of ice as often as every five weeks. What happens next depends on the currents that drive them, on the waves that slap and pound at their flanks, and on a number of other forces that glaciologists still do not fully comprehend.

The largest and most challenging icebergs calve off the continent of Antarctica. There, no less than 7,000 miles, or 30 per cent, of the entire coastline consists of sheets of glacial ice that have thrust their outer edges into the sea without detaching themselves from the land behind. The shelves are located in broad coastal embayments, where currents are gentle, the tidal range is slight, and pack ice—the frozen sea water that forms in plates seven to 12 feet thick along the shore—cushions them against wave turbulence and ocean swells. The shelves press seaward at speeds that range from three to eight feet a day, until they reach massive proportions. The largest of them, the Ross Ice Shelf, which lies between McMurdo Sound and Little America, is more than 500 miles wide at its terminus, where it forms a sheer cliff that towers as high as 150 feet above sea level.

The farther out to sea the ice shelves reach, the more pronounced are the currents, tides, waves and swells they are apt to encounter. At some critical point when a shelf has become too big to withstand the stresses of ocean forces, it calves, breaking off great slabs. Most of the icebergs are relatively smooth on top, for the snow line on these shelves is at sea level, and repeated snowfall during the years of the glacier's growth has filled in the depressions and leveled off the rises that might have formed during the glacier's movement. In length these icebergs average a mile or two when newly calved, but icebergs five miles long are common—and some are much bigger than that: In 1956 a behemoth more than 208 miles long and 60 miles wide was sighted.

How much ice is calved in all during an ordinary year in Antarctica is anyone's guess, for no routine patrolling is done in this little-traveled region, and scientists' estimates depend on extrapolations of precipitation, evaporation and spot counts of limited areas. One such spot count was made in 1965, and some 30,000 icebergs were found to be lying between long. 44° and 168° E., an area of 1,700 miles.

Once loosed from the ice shelf, Antarctic icebergs move away from the coast in a northerly direction, simultaneously drifting from east to west because the general pattern of ocean circulation near the continent follows the prevailing geostrophic winds—those that are caused by the rotation of the earth on its axis. The drifting icebergs have been clocked at speeds of seven or eight miles a day. But their movement is not constant; the icebergs may make many detours and stopovers as they encounter eddies that whirl them in circles, shoals that beach them, bays and inlets that trap them.

Many of them will become permanently stalled. Of these, most succumb to the ravages of waves, winds and solar radiation; but along a short stretch of the continent at about 80° E., icebergs may actually increase in size. The easterly winds here drive them inshore, where they are reincorporated into the ice shelf, to be calved into the ocean once more at some future date.

Icebergs that escape into the open sea eventually meet the so-called Antarctic convergence, or Circumpolar Current, which meanders through the Pacific, Atlantic and Indian Oceans between lat. 65° and 50° S. Here they reverse course and move from west to east. Once in a while an iceberg will be hurled out of that current by an eddy and sent farther north; in 1892 and 1967, two particularly stormy years, icebergs were sighted at about 45° S. —the latitude of New Zealand—and in 1895 one was sighted at lat. 40° off the coast of Africa. But for most Antarctic icebergs the Circumpolar Current represents the outer limits of their journey. Its course is a swift one knot, its temperatures hover around a warm 40° F., and no pack ice remains to buffer the bergs. It is doubtful that any of them even complete the 10,000-mile circuit. No one can say with any precision how long one of these icebergs will survive. The great variation in their size and the temperatures, currents and storms they encounter all combine to lengthen one iceberg's life span and shorten another's.

A series of photographs details the calving of icebergs into a fjord from Greenland's Daugaard-Jensen Glacier in 1968. Photo sequences like this one showed that the faster-moving center of the glacier invariably calved first.

All that can be said for certain is that by the time they reach 65° S., they have already shrunk to a mere fraction of their original majestic size; bergs that were once a mile long will be no more than 1,300 feet. The process of disintegration steadily accelerates, since the smaller an iceberg becomes, the more vulnerable it is to the erosive forces of heat and turbulence, for reasons that have to do with the changing ratio between mass and surface area. Anybody who has ever tried to unstick a jumble of ice cubes in an ice bucket has seen the phenomenon at work. The jumble will remain as is for hours, but an equal number of ice cubes lying separately in a sink will melt in minutes. That is because a dozen separate one-inch cubes have masses of one cubic inch apiece, and 72 small surfaces simultaneously exposed to the effects of warming air. The same cubes jumbled have most of their surfaces turned inward—where they unite to enlarge the mass, and all help to keep one another cool. By the same principle, the fragments of a disintegrated iceberg melt faster than did the parent iceberg.

For years it was widely believed that outsized tabular icebergs were unique to the Antarctic and not to be found in the Northern Hemisphere. But in August 1946, observers aboard a U.S. Air Force weather plane cruising over the Arctic Ocean learned otherwise. The crew could scarcely credit their eyes when the radarscope picked up a strange heart-shaped mass about 300 miles north of Point Barrow, Alaska. At first, because of its size and density, they took it to be an island. But the nearest known island was some 500 miles distant. For lack of a better name, the fliers called it Target X. Subsequent inspection showed that it was moving through the pack ice, which proved that it could not be land. Indeed, it was finally found to be a mammoth slab of ice 18 miles long and 15 miles wide. It was estimated to have a total thickness of about 100 feet, approximately 1/10 the depth of the thickest Antarctic bergs. Otherwise it was essentially no different from the great tabular icebergs found off Antarctica. Four years later, in July 1950, two more such giants were discovered in the Arctic Ocean. Target X was renamed T-1, and its successors came to be designated T-2 and T-3.

The Air Force tracked T-1 for three years, and watched it travel northward a distance of 1,400 miles. Then it was lost somewhere in the polar ice pack. After another 22 months had passed, it was sighted again, this time near the north coast of Canada's Ellesmere Island. Resuming the tracking, the Air Force watched it turn to the west and move along the northern edges of the Canadian archipelago in the direction of Point Barrow—where it had first been spotted five years before.

Meanwhile, the startling discoveries had prompted scientists to re-examine aerial photographs made over the years, and suddenly it was obvious that a goodly number of these "ice islands" existed. They were distinguishable because their unbroken expanse stood out from the shattered ice pack around them. After the initial surprise wore off, scientists realized that the curious islands no doubt account for some of the strange sightings reported by 19th Century explorers. Many an Arctic explorer avowed that he had spied land near the Pole, only to have his successors find nothing but floating pack ice. Scientists and historians had written off the reports as indications of hallucination or, most charitably, as gross inaccuracy of charting.

In the eight years following the first discovery, 50 or so of the ghostly islands were identified floating around the North Pole. And scientists later estimated that perhaps 100 all told were adrift in the Arctic Ocean. On examination, the configurations of the great tabular icebergs showed that most of them had calved off Ellesmere Island (which has five ice shelves that extend up to 30 miles into the Arctic Ocean) and perhaps off one or two other glaciated islands of the Canadian archipelago. Some of the iceberg edges fitted like jigsaw pieces into the edges of the Ellesmere shelves. The surfaces of all of the icebergs resembled those of the shelves—essentially flat but rippled with low ridges and shallow troughs formed by meltwater. And some were laden with glacier-borne boulders, clay, moss and twigs swept down from mountain valleys in the interior. Comparison of the rings of willow twigs found on T-3 with other willows remaining on Ellesmere Island made dating possible, and scientists estimated that T-3 had been attached to land as late as 1935 before embarking on its Arctic odyssey.

Their mobility, in combination with their size and stability, made these tremendous bergs highly useful to science. Among them were several that were long enough to accommodate aircraft, durable enough to be habitable and wide-ranging enough to give researchers in half a dozen disciplines a grand tour of the North Polar region. In 1952 American scientists occupied T-3 and turned it into a floating laboratory. The Russians, who have 14,000 miles of Arctic coastline, later occupied another of the ice islands.

One of the most interesting findings to emerge from the ice-borne laboratories was that the Arctic Ocean seabed, which had previously been thought to consist of a single large basin, is in fact a complex grid of plains, plateaus, escarpments, abysses—and three mighty submarine ridges that run parallel to one another from north of Greenland and the Canadian archipelago toward Siberia. The most majestic of them, the Lomonosov Ridge, rises an average 10,000 feet above the ocean floor and, at its highest points, to within 3,000 feet of the ocean surface.

The scientists also found that the ocean water circulates over this varied seascape in two main currents. One flows in a broad arc from the Chukchi Sea, north of Alaska, and the Bering Strait across the Asian side of the ocean and around toward the Greenland Sea. The other is a clockwise gyre in the Canadian side of the ocean.

The latter is the one that ice islands follow. In all, they make a 2,300-mile loop in the ocean just north of the North American continent. Starting from the north shore of Ellesmere Island, the ice islands float with the current along the edges of the Canadian archipelago, then swing north above Alaska. Somewhere short of the Pole they turn south and return to their starting point on the shore of Ellesmere Island. At an average speed of a little more than a mile a day, an ice island takes 10 to 12 years to travel the full route—provided it keeps moving. Some bergs proceed fitfully, as T-3 has shown. Two years after oceanographers and meteorologists landed on that great berg, when it was near the North Pole, T-3 returned to the edge of Ellesmere Island, where the scientists abandoned it. In later years it alternately ran aground and drifted free. At the start of the 1980s it was on the move again in the Arctic Ocean, transporting no human passengers, but carrying an electronic beacon that automatically reported its position to a satellite. Assuming that scientists had calculated correctly when they reported that it was first detached from Ellesmere Island in 1935, the mammoth iceberg was 45 years old in 1980.

The all-encompassing pack ice accounts for the longevity of Arctic bergs. But not all ice islands are as long-lived as T-3. The current they ride travels not in a true circle but in a gyre; it widens with every circuit and may ultimately cast them out of the protective pack. A case in point was an ice island named Arlis II (for Arctic Research Laboratory Ice Station No. 2). Measuring two by three miles, it served American scientists as a floating laboratory for four years, beginning in 1961, and promised to be useful for many years to come. But when Arlis veered west in the Arctic Ocean in late

1963, it did not return to Ellesmere Island; instead, it was somehow dumped out of the Arctic gyre and into the Greenland Sea. There it was picked up by the southward-running East Greenland Current, which carried it down along that country's coast and into warmer waters, sometimes at the astonishing rate of 24 miles a day. By May 1965 the pack ice around Arlis began to thin out, the island was threatening to split right down the middle of the research camp, and the researchers were forced to evacuate.

They left behind two radio beacons that allowed them to follow Arlis' path. The island sailed on into the Atlantic southwest of Iceland. By May 31 the expected breakup had begun. One calving followed another, and by June 20 Arlis II had been reduced to about a dozen fragments. These shards of the original berg had rounded Cape Farewell at the southern tip of Greenland by the end of July 1965. When last seen, they were riding the northward-flowing West Greenland Current toward Baffin Bay—and eventual extinction among the other disintegrating icebergs, growlers and brash crowding into that body of water.

Baffin Bay, which laps 600 miles of the frigid shores of Greenland, contains perhaps the greatest concentration of icebergs in the Northern Hemisphere. Greenland differs from the Canadian archipelago in that it has no ice shelf. Instead, its primary glacial feature is a monumental 665,000-square-mile interior ice sheet that flows seaward in the form of more than 16 glaciers. The most prolific of these glaciers is the Jakobshavn, which flows at a rate of about 75 feet a day during July and August, and disgorges 20 to 30 million tons of ice daily into the fjord at its snout. Altogether Greenland's glaciers keep the waters surrounding the island choked with an estimated 20,000 icebergs year in and year out.

Icebergs that calve on the east coast move south toward the Atlantic Ocean on the East Greenland Current, following the same route taken by Arlis II when it escaped from the Arctic Ocean. If they survive the passage around Cape Farewell, they head north into Baffin Bay. Few are massive enough to do so; most of them break up into bergy bits and growlers, just as Arlis II did. But the west coast, having the nine major glaciers of the island, sends its own icebergs plunging into Baffin Bay—a total of nine trillion tons of them annually.

The western glaciers take two forms, depending on which half of the

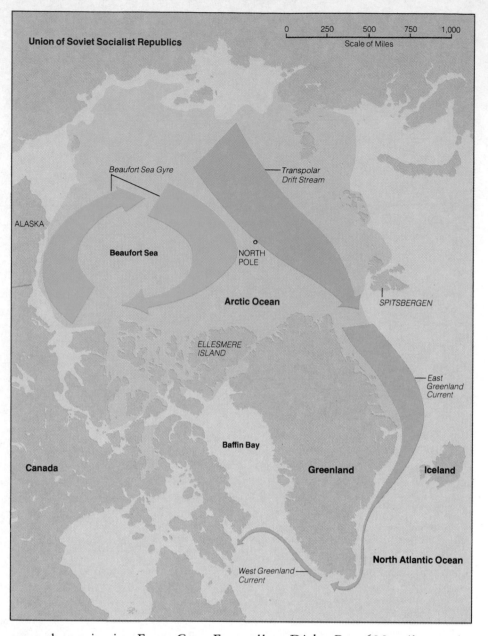

The major currents indicated on this map
tend to shepherd icebergs along regular routes
through the northern oceans. From the ice
shelves of Canada's Ellesmere Island, icebergs
trace a clockwise loop, called the Beaufort
Sea Gyre, in the Arctic Ocean. Icebergs
discharged along Greenland's coast follow
the offshore currents into Baffin Bay, where
they are picked up by southerly currents.

coast they arise in. From Cape Farewell to Disko Bay 600 miles to the
north, the glaciers terminate in fjords, which sometimes stretch more than
100 miles inland from the sea. Their icebergs accumulate on shallow underwater ridges that temporarily halt their advance to the sea. About twice
a month, when full moon and new moon bring exceptionally high tides,
the surge of water dislodges the backed-up ice, and the bergs—at the
Jakobshavn, about 300 million tons of them—lurch free and rush for open
water. The sounds echoing back along the walls of the fjords resemble those
of a battlefield, with cracking rifle fire and the low rumble of artillery,
and the waves churned up can capsize small boats.

North of Disko Bay the Greenland coastline straightens out, the fjords
disappear and the glaciers flow majestically to the very edge of the sea.
Some of them stand as high as 150 feet above sea level, and a number reach
as far as four miles into Baffin Bay. These massive glacial snouts are constantly discharging ice into the sea—sometimes as monumental tabular
bergs that are several miles long and rise 200 feet above the water; sometimes in the form of bergy bits, growlers and brash; sometimes as massive
sheets that plunge from the full height of the glacier.

The Eskimos who lived along one such Greenland glacier understandably viewed it with superstitious reverence. They believed that if they talked, laughed, smoked or spoke the glacier's name when paddling past, it would take offense and angrily smother their boat with a shower of punishing ice. A peal of laughter is not likely to provoke a glacier, but it is best not to overdo it, for there is some truth to the Eskimo belief. So precarious is the riven and melting front of a glacier that any substantial atmospheric disturbance may cause a sheet of ice to lose its purchase and go plunging into the sea—as 19th Century explorers occasionally discovered to their sorrow when they discharged muskets at walrus cavorting under the walls of Greenland glaciers.

Even at a safe distance, the spectacle of a calving glacier is awesome. After witnessing a great chunk of ice tumble from a Greenland glacier in 1818, British explorer F. W. Beechey wrote in wonder: "The piece that had been disengaged at first wholly disappeared under water, and nothing was seen but a violent boiling of the sea, and shooting up of clouds of spray, like that which occurs at the foot of a great cataract. After a short time it reappeared, raising its head full a hundred feet above the surface, with water pouring down from all parts of it; and then, laboring as if doubtful which

Icebergs afloat in Newfoundland's St. Lunaire Bay in June 1974 provide graphic evidence of the severity of that year's infestation; icebergs usually melt in these waters by April or May.

way it should fall, it rolled over and, after rocking about some minutes, at length became settled."

From among these Greenland icebergs come the ones that have brought epic human misery to the North Atlantic shipping lanes 2,000 miles to the south. The icebergs travel a circuitous route to get there. In Baffin Bay they ride north on the West Greenland Current to about the 75th parallel, then arc around the bay to come south again on a series of currents past Baffin Island and Labrador. All the while their numbers are diminishing, but an average of 400 bergs annually survive the journey as far as the 48th parallel, where they arrive at the Grand Banks, the shallow continental shelf that extends into the North Atlantic off the southeast coast of Newfoundland.

For most icebergs the area of the Grand Banks marks the end of the journey. Here they meet the Gulf Stream as it turns eastward. That powerful current, with its 50° to 60° F. temperatures, generally melts them in two weeks. But in an unusually severe year, 35 or 40 hardy icebergs may be shunted from the Gulf Stream into some cold-water meander that flings them farther south. If they pass below the 48th parallel they enter the shipping lanes, where icebergs have plagued mariners since the dawn of transatlantic seafaring.

In the early days, the disasters wrought by icebergs were regarded as acts of God—to be feared but suffered with stoicism. This fatalistic attitude persisted into the late 19th Century, when the invention of the steamship increased passenger traffic dramatically and the loss of life became appalling. Between 1882 and 1890 alone, no fewer than 14 liners were sunk and another 40 were damaged by ice near the Grand Banks. Dozens of other vessels went missing in the area, many of them undoubtedly claimed by icebergs.

But in 1912 one terrible iceberg that traveled south to the 42nd parallel demonstrated the enormity of the menace and galvanized a number of seafaring nations into action. On April 10 of that year the world's newest and greatest liner, a 46,000-ton beauty with a gleaming black hull and four buff-colored funnels, left Southampton, England, bound for New York on her maiden voyage. In size, in design, and even in name —the *Titanic,* after the Greek gods of earth and ocean—she suggested invincibility. She was 882.9 feet long, possessed a double bottom of three-quarter-inch steel plates, and was partitioned belowdecks into 16 compartments that could be made individually watertight; a push of a button on the bridge would close the doors between them. She carried with her the greatest of expectations.

On board for the gala crossing were 2,206 passengers and crew. The ship's list was graced by a glittering array of wealthy and titled personages, among them financier John Jacob Astor, Macy's department store owner Isidor Strauss, several members of the British nobility and Thomas Andrews, the proud builder, who looked upon his creation as the epitome of 20th Century progress. The crewmen shared in the sense of pride. "God himself could not sink this ship," a deck hand was heard to say as the *Titanic* glided out of Southampton harbor.

Such complacency was not to survive the voyage. At 11:40 on the cold, starry night of Sunday, April 14, as the *Titanic* was knifing through the sea southeast of Newfoundland at a brisk 22.5 knots, the lookout in the crow's-nest sighted an iceberg dead ahead, looming out of the darkness like some fantastic wraith. He immediately rang the alarm bell and telephoned the bridge to report his sighting. The duty officer swiftly ordered a halt to all engines and a turn hard astarboard; as a precaution, he initiated the closing of watertight compartments belowdecks.

Yet quick as he was, the orders came too late; scarcely 30 seconds after the lookout's report, the duty officer heard a terrible grinding noise. The iceberg was scraping the side of the ship, and below him he could see tons of ice, from small slivers to chunks the size of basketballs, cascade onto the starboard well deck opposite the foremast.

Incredibly, even in that deluge of ice, no one sensed the dimensions of the calamity. Passengers picked up pieces of ice and began playfully pelting one another; someone in the second-class smoking room, hearing of the sport, jokingly asked if he could have a few bits of the ice for his highball. Some cardplayers looked up from their table to see the berg sail past a porthole and then, undisturbed, returned to their game. Not even the shipmaster was alarmed; Captain E. J. Smith, a veteran of 34 years at sea, presently began a tour of the ship with builder Thomas Andrews, but made no move to lower the lifeboats, rouse off-duty officers or radio a distress signal. The damage was invisible—and more terrible than he could have imagined. The berg had sheared off the hull's rivets and wrenched a 300-foot opening in the steel plates. Within 10 minutes water would rise 14 feet above keel level, flood five of the 16 compartments, and begin to seep under the doors of the crew's quarters and the steerage cabins on the lowest deck.

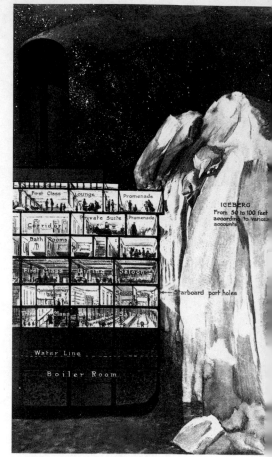

The ill-starred *Titanic* collides with an Arctic iceberg in this illustration from a 1912 British periodical. The cutaway drawing shows why the passengers thought the iceberg had merely brushed the great ship: The 300-foot gash that would send her to the bottom with the loss of 1,503 lives was in her side below the water line.

As he became dimly aware of the impending catastrophe, the captain ordered the crew to man the pumps, but not for half an hour did he perceive the grim truth that with water rushing in at a fantastic rate, pumping was hopeless. Belatedly Captain Smith ordered the lifeboats lowered and a distress call sent out to other ships at sea. In the confusion that followed, only 703 passengers—not even a third of those on board—reached the safety of the boats, while the ship listed crazily and sank inch by inch into the frigid North Atlantic. Finally, at 2:15, almost two and a half hours after the collision, the sea washed over the bow of the *Titanic*. Her stern rose into the air, and she stood perpendicular for a moment as her innards—engines and elevators, furniture and food, crystal and coal—broke loose and went tumbling forward with a shattering din. Then came an eerie silence as the magnificent liner slid to her grave, carrying 1,503 human beings with her.

Nearly two hours passed before the first rescue vessel arrived on the scene. Her crew found the *Titanic's* lifeboats scattered over four square miles of iceberg-strewn waters, and in the first gray light of dawn the rescuers could scarcely distinguish the boats from the ice.

The berg that had destroyed the world's proudest ship was never identified for certain. But later on the morning of the *Titanic's* demise the chief steward of a German ship photographed an iceberg that bore a great livid scar along its base—presumably left by the red paint on the water line of the *Titanic*. That iceberg was about 80 feet high and 60 feet long—only medium-sized as Grand Banks icebergs go—but its estimated displacement was 200,000 tons, four times that of the *Titanic*. And that enormous weight combined with the force of the speeding vessel to cleave the *Titanic's* steel hull in a trice.

The magnitude of the *Titanic* tragedy, with its colossal loss of life, triggered a great public clamor for safety measures against icebergs in the sea-lanes. Within a week the United States Navy dispatched two small cruisers to the Grand Banks near where the *Titanic* had gone down, with orders to patrol for icebergs, chart their positions and broadcast the coordinates to ships at sea. The following year two ships from the Revenue Cutter Service—forerunner of the Coast Guard—took over the patrol duty. On January 30, 1914, representatives of 13 maritime nations met in London and agreed to share the responsibility. They authorized the creation of the body now known as the International Ice Patrol, to be funded by the signatories and administered by the United States Coast Guard.

Every year since then, except during World War II, units of the U.S. Coast Guard have scoured a 370,000-square mile area around the Grand Banks—an area known to the men as Iceberg Alley—in the service of international shipping. From the 52nd parallel south to the 42nd and from long. 40° to 50° W., guardsmen monitor icebergs, counting the number that enter the area, measuring their size and charting their drift. The patrol issues radio bulletins of its findings twice daily from stations in the United States, Canada and Europe for the benefit of ships that ply the route. Moreover, glaciologists and oceanographers frequently sail with the patrol in order to expand scientific understanding of icebergs and how they behave; indeed, the International Ice Patrol is the source of much of today's knowledge of icebergs.

Finding and tracking potentially dangerous icebergs in a vast expanse of sea can be a daunting task. In the first days of the Ice Patrol, the only way was to position sharp-eyed lookouts aloft in the crow's-nest of a cutter. But the Grand Banks, lying at the juncture of the cold Labrador Current and the warm Gulf Stream, are notorious for fog, and from May through July visibility is restricted to a few hundred feet perhaps 80 per cent of the time. In such cases, the cutter captain might sound the ship's whistle, in hopes that an echo would enable him to locate a nearby berg. But that could be dangerous. As occurs along glacial ice shelves, the sudden atmospheric vibrations might cause an unstable berg to calve, and a huge chunk of ice breaking off into the sea would send towering waves crashing at the cutter.

After World War II the Ice Patrol adopted aerial reconnaissance. Today, a single aircraft—the C-130 turboprop transport, equipped with radar—can take off from Gander, Newfoundland, and cover an area of 30,000 square miles in a flight of six to eight hours, depending on weather conditions. Flying in good weather at 1,500 feet, a pair of experienced observers

On its annual mission for the International Ice Patrol, the U.S. Coast Guard cutter *Evergreen* tracks an iceberg off Greenland's west coast. Oceanographers on the *Evergreen* study long-term patterns of iceberg drift and deterioration.

manning stations on board a C-130 can spot icebergs at a range of about 12 miles on either side of the flight path. In poor weather, the planes drop down to as low as 500 feet in search of their quarry.

Even in the best of conditions the human eye is hard put to keep track of icebergs because they are such changelings. An iceberg sighted today may well exhibit a totally different face tomorrow. In 1976 a group of glaciologists was watching a 6,000-by-2,000-foot tabular iceberg in a Greenland fjord when for no apparent reason it suddenly rolled over, shedding layers of fissured ice in the process, and then rolled back again. No sooner had it reached equilibrium than it split in two and was transformed beyond recognition.

The Ice Patrol has experimented with a wide variety of identification schemes for dangerous icebergs. The problem is complicated by the fact that the bergs posing the gravest threat—the pinnacled, unstable ones—are also the most elusive. They offer few flat planes on which to attach identification, and even a plane that is exposed one moment may be submerged the next. For a time, the Ice Patrol splashed potentially troublesome bergs with bright red, green and blue dyes. But the icebergs shed their coats of color as fast as they melted, and obliterated all evidence of marking when they rolled. Another method was to cinch an iceberg with a floatable collar containing radar reflectors and a radio transmitter. But again melting altered the icebergs' contours, and they slipped out of their collars.

An ingenious new system was devised in 1979, when Ice Patrol researchers began to make use of steel darts and radio transmitters in conjunction with the TIROS weather satellite in polar orbit more than 500 miles above the earth. A steel dart nearly four feet long was dropped on an iceberg from a low-flying aircraft. Attached to the dart was a 984-foot-

A crewman aboard the *Evergreen* takes aim at an iceberg with a dye-tipped arrow. Splotching bergs with bright colors to tag them for future study was abandoned in the mid-1960s by the International Ice Patrol because melting and wave action promptly washed the dyes away.

long tether made of polypropylene, which floats. At the other end of the tether was a radio-equipped buoy that was dropped into the water by parachute. Wherever the iceberg drifted, the researchers reasoned, the transmitter broadcasting to the satellite would follow like a noisy can tied to the tail of a dog.

The system succeeded well enough for scientists to follow the wanderings of one or two bergs. But even this method had its limitations. If the iceberg calved, the dart might spring free and sink into the deep. If it remained attached to the ice and continued transmitting, there was no telling by satellite whether the message was from an iceberg that remained a peril or from a harmless piece of brash. Moreover, the system proved far too costly for tracking the hundreds of icebergs that pass into the Grand Banks area each year. Its value thus seemed greatest for research into general drift patterns—not for large-scale surveillance.

In any event, tracking alone is not sufficient to deal with the menace that icebergs present. Since 1968, oil and natural gas have been discovered in Arctic regions—in the North Slope of Alaska; on the edges of Baffin Bay and the Labrador Sea; off Newfoundland, virtually in the middle of the Grand Banks; and in the Canadian archipelago. In all of these locations, icebergs pose a variety of threats to the industry. An iceberg that looms 100 feet above the water line extends 700 feet below it—and will rip up anything it encounters on the ocean floor, including a gas pipeline. An iceberg could crumple a $50-million oil rig or a tanker as quickly as it sheared the side of the *Titanic,* causing a monumental oil spill with all its dread effects.

From its inception, the International Ice Patrol has experimented with ways of actively ridding the sea-lanes of icebergs instead of merely reporting their presence. If a berg could be blown apart, it would melt much more rapidly. But icebergs disintegrate on nature's timetable, not man's, and they stubbornly resist all attempts to speed up the process. "For diversion and in the nature of an experiment we fired 12 six-pounder shots at an iceberg," wrote the captain of a cutter patrolling the Grand Banks in 1914, "but they were just as effective as if we had attempted to storm the Rock of Gibraltar."

Since then the Ice Patrol has assaulted icebergs with torpedoes, mines, depth charges and bombs—with similar lack of effect. "Over the years we've tried everything," former Ice Patrol Commander A. D. Super once said. "You can't detonate or destroy them. We tried and we just made little dents." It has been estimated that to melt a 500,000-ton iceberg would require the heat produced by burning one million gallons of gasoline.

Since they cannot get rid of the icebergs, engineers are trying to find ways of safeguarding their operations. One promising approach is to house drilling rigs in mobile platforms that could be moved to safety when threatened by an iceberg—but to disconnect the cumbersome platforms for removal may take as long as a day and a half. Another method, rapidly gaining favor, is to tow the icebergs themselves aside with powerful tug-

An explosive charge set by the crew of a U.S. Coast Guard cutter in 1945 spews ice and water 200 feet above an iceberg off southeast Greenland. The iceberg, which was blocking a shipping channel, was only nicked by the spectacular explosion; a second charge sheared off enough ice to let the cutter pass.

boats hitched to a harness of four-to-12-inch-thick hawsers made of braided polypropylene.

Towing is difficult work at best, and hauling an iceberg can be hazardous in the extreme. An iceberg in tow is just as likely to calve or flip over as one riding free—and any tugs in close proximity could be swamped by mountainous waves. Moreover, the mass and momentum of icebergs are so great that fantastic applications of power are required to produce any effect. Deep-water currents, which often flow in different directions from those on the surface, may work at cross-purposes with the towboat, tugging the deep-draft iceberg in the wrong direction. Curves and ridges in the underbody of the iceberg may provide a ruddering effect and steer it awry. "You never know whether you are towing the berg, or whether it is towing you," says one man who took part in a towing operation. Still, for the purposes of protecting an oil rig, there is no need to tow an iceberg any great distance; it is sufficient to deflect it a few hundred feet from its path of destruction. Towboats have proved themselves capable of doing the job. They represent the first successful attempt by man to curb the power of icebergs and strip them of their menace.

For all their baneful traits, icebergs have also been viewed as a potential boon to mankind. Their one obvious asset is fresh water, and that is a commodity in scant supply in much of the world. In the 18th Century, when the pioneering navigator Captain James Cook led two sailing ships on a voyage of discovery to the South Pacific and crossed the Antarctic Circle, he took on 25 tons of iceberg fragments to replenish his failing water supply. Cook declared the operation "the most expeditious way of watering I ever met with," and other mariners in later years slaked their thirst and kept themselves alive by coming across icebergs.

Boundless fresh water from the Poles remains a tantalizing dream. Antarctica alone has locked up in its slowly melting fleets of icebergs the equivalent of one third as much water as the entire world annually consumes. What is more, the water is the purest imaginable. Because Antarctica lies two thirds of a world away from the industrialized Northern Hemisphere, its glacial waters contain scarcely one part per billion of organic matter. To

Harnessing an Iceberg for War

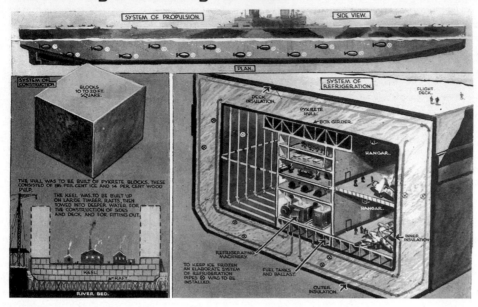

Cross-sectional drawings for the British aircraft carrier *Habbakuk* show how a thick hull of the ice-and-wood mixture Pykrete was intended to house hangars, crew quarters, a refrigeration system to prevent melting, and an electric generator to drive the ship's 26 motors.

In the desperate early days of World War II, when Great Britain needed more aircraft carriers to launch planes against German submarines in the Atlantic, an inventor named Geoffrey Pyke began musing on icebergs. He knew that they were virtually impervious to explosives, and that Arctic pilots had sometimes used large ice floes as airstrips. Would it be possible, he wondered, to construct inexpensive aircraft carriers of ice?

Pyke's idea for a "bergship" won the enthusiastic support of Prime Minister Winston Churchill in 1942, and the inventor drew up plans for H.M.S. *Habbakuk*, a carrier appropriately named for the Biblical prophet who promised "a work in your days, which ye will not believe, though it be told you."

The design called for a hull 2,000 feet long that would displace 1.8 million tons of water—26 times the displacement of the liner *Queen Elizabeth*. Its 50-foot-thick walls would be built of Pykrete (Pyke's concrete), a frozen solution of water and wood pulp that was far stronger than regular ice because its wood fibers reduced the tendency of the ice crystals to slide over one another.

When a 60-foot model of the *Habbakuk* was built at Patricia Lake in Alberta, however, the British began to have second thoughts. Even in Canada's sub-zero temperatures, it would take 8,000 men eight months to freeze and assemble the nearly 280,000 Pykrete blocks required for the full-sized ship. At an estimated cost of $70 million, the *Habbakuk* would be just as expensive as a conventional steel carrier—but far more problematic. Thus, in late 1943, after the Allies' antisubmarine campaign had succeeded without the *Habbakuk*, Pyke's bergship project was scuttled.

leaders of arid nations, where an estimated one billion people are in dire want of water, the idea of all that potentially fresh, pure water going to waste is profoundly frustrating.

Some serious consideration of the problem got under way in California in the 1950s after a searing drought had left that state desperately parched. Oceanographer John Isaacs at the Scripps Institution in La Jolla proposed in a scholarly seminar that Antarctic icebergs be towed to California. He chose faraway Antarctica instead of Alaska because the large tabular icebergs there are more stable than their smaller, pinnacled kin in the Arctic. He calculated that a fleet of six oceangoing tugboats could bring a 20-mile-long iceberg to California from below lat. 65° S. in a voyage of about six months.

"We thought it was a dumb idea," recalled Wilford Weeks, a glaciologist at the U.S. Army's Cold Regions Research and Engineering Laboratory in New Hampshire, "so we planned to ice it once and for all by writing a technical paper about it." To his own astonishment, Weeks ended up by concluding that towing Antarctic icebergs was feasible after all. He thought that California might be out of range, but he saw real possibilities for nearer destinations, such as the west coasts of Australia and South America, that could be reached without passing through the tropical waters of the Equator.

In 1973, another team of scientists published an even more enthusiastic endorsement of iceberg towing. They were John Hult and Neil Ostrander, two Rand Corporation scientists charged with finding some practical use for the information gleaned from satellites orbiting over Antarctica. Musing over all those icebergs, they came up with a scheme for transporting them, not just singly, but eight at a time, linked together by hawsers like a train of barges. A nuclear-powered towboat would push from behind, electrically powered propellers would be harnessed to the sides of the icebergs in the train, the prow of the lead berg would be carved in the shape of a ship's bow to make the whole slip easily through the water—and a few escort vessels would chug alongside this concoction to make sure that all went well.

Hult and Ostrander were so smitten with the prospect that they resigned from Rand and formed their own company to transport ice to California. They scaled down their initial idea of an iceberg train and began work on a pilot project of towing one or two modest-sized bergs no longer than 1,100 feet each from Antarctica to Southern California. The California legislature endorsed the scheme but supplied no funds, and that proved fatal. It would have cost $30 million to put the pilot project into operation, and no private investor or combination of investors could be persuaded to risk so much capital on so radical an idea. A science writer summed up the problem nicely when he observed that to bring off such a project "requires an unusual combination of wealth and thirst."

But just when it appeared that iceberg towing would remain in the realm of science fiction, a man representing that very "unusual combination" stepped on stage. He was Prince Mohammed al Faisal, a son of the late King Faisal of Saudi Arabia, and in his own right a banker, entrepreneur and government leader with a personal fortune reckoned to be in the tens of millions. His native land was exceedingly thirsty; Saudi Arabia has scarcely any rivers or lakes, and averages just four inches of rainfall annually. The need for water was so critical that by the 1960s the Saudis had begun a gigantic sea-water desalinization project that was expected to cost an astronomical $15 billion. Faisal himself was chief of the program—and so ardent a champion did he become that he was known as the Prince of Water. Now Faisal thought he had found a far better answer in the paper written by Hult and Ostrander. According to the American scientists, fresh water could be processed from icebergs more cheaply than from the sea—perhaps for little more than half the cost.

In 1977, the Prince formed a company for the sole purpose of finding a way to transport a 100-million-ton iceberg from Antarctica to the Saudi port of Jidda on the Red Sea—a distance of some 9,000 miles. In the first year, Faisal and his company spent an estimated one million dollars just gathering information. He sponsored several conferences to pick the brains of scientists and engineers from all over the world. At one such conference, in Ames, Iowa, in 1977, the Saudi Prince made a dramatic show of chilling drinks with ice chipped off a 4,785-pound growler that, at a cost of $14,000, he had had plucked by helicopter from Alaska, ferried to Anchorage, packed in a huge Styrofoam box, flown to Minneapolis, trucked to Ames, and unloaded on the campus of Iowa State University. The fanfare did not get iceberg transport under way, but the conference did generate ideas on how it might be done.

The first problem is how to propel icebergs thousands of miles through tropical waters without their melting away. The most likely means of pro-

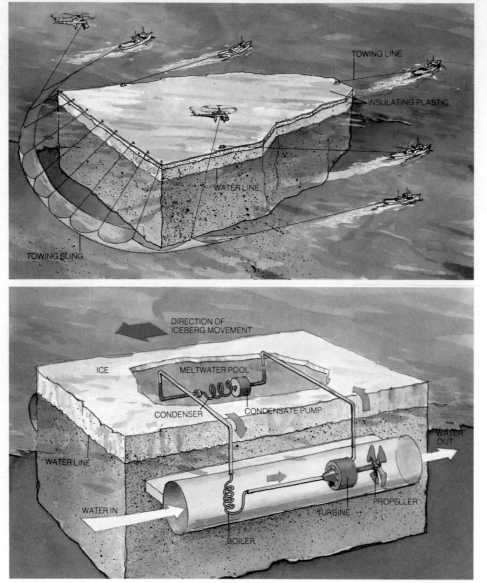

One scheme for utilizing the fresh water locked in a mile-long tabular iceberg from Antarctica envisions swaddling the berg with layers of plastic insulation to protect it from the sun and warm seas. After tugboats and helicopters had positioned a huge sling around the iceberg's underwater bulk (*left*), the immense package would be towed to thirsty lands far to the north.

Another plan for iceberg transport calls for attaching a propulsion pod to the side of the berg to make it self-propelled. Warm sea water would be used to evaporate liquid ammonia under pressure in the boiler. The pressurized ammonia vapor would rush through the turbine to drive the propeller, then cool to a liquid once more in the condenser on top of the berg for reuse in the boiler.

pulsion would be a fleet of half a dozen or so conventional oceangoing tugboats. They could be deployed in a single, evenly spaced rank about 800 yards in front of the berg. Their towlines would be affixed to a harness surrounding the iceberg.

The tugs would have to be the largest and most powerful kind in existence, with 20,000 horsepower. Even so, the inertia of a 100-million-ton weight is such that the fleet would require as long as 24 hours to reach a speed of one knot—slightly more than one mile per hour. At that rate, the journey from Antarctica to Saudi Arabia would require a year. But to go faster would greatly increase the frictional drag of the berg, thereby hastening its melting. Even traveling at such a slow speed an iceberg would not be likely to make it past the Equator without some form of insulation. "You would end up with nothing but a towline," said Army glaciologist Wilford Weeks in 1980.

Insulating the iceberg is a nettlesome challenge that has inspired a variety of suggestions, from spraying the entire berg with plastic foam to wrapping it in canvas. John Hult, the Rand engineer turned iceberg aficionado, has suggested a kind of plastic swaddling to insulate the berg. It would consist essentially of two layers of plastic sheets separated a few inches from each other by a layer of air-filled tubes and rigged to catch the cold meltwater, which would help shield the iceberg from the debilitating warmth of the sea water. Hult contends that this system would limit melting to only 5 per cent in a year.

Neither that idea nor any other proposed for insulation appears to solve a problem that may be even more serious than melting: the proclivity of icebergs to calve. Calving en route would not only greatly reduce the volume of ice reaching its destination, but might unleash dangerous growlers and bergy bits to menace the shipping lanes, which in the Southern Hemisphere extend as far south as the 30th parallel.

If an iceberg should reach its destination reasonably intact, there remains the problem of what to do with it then. No harbor in the world can take a berg with a 750-foot draft, which means that the ice would have to be anchored many miles offshore. In the case of Saudi Arabia, the berg could not even enter the shallow mouth of the Red Sea. One suggestion is to station the iceberg a few miles offshore, and then slice it up like a cake by means of thermoelectric wires and tow the pieces separately into port.

Not much thought has yet been given to the payoff for all this effort: actually harvesting water from the ice. Pools of meltwater on top of the iceberg might be pumped ashore in a pipeline, or the ice might be broken up like coal and the small chunks sent through pipes in a slurry of water. All the while, however, the ice below the water line would be melting and mixing with salty sea water, and going to waste—until and unless some way is devised for keeping the meltwater pure.

Finally, there are questions about the effects on the environment: Will all that ice in a warm climate create a dense bank of fog overhead? Will it influence wind and ocean currents? Will it destroy marine life that depends upon warm water?

Many scientists at the conferences concluded that Faisal ought to begin small by undertaking a less ambitious journey than the one to Saudi Arabia—perhaps a 2,000-mile tow from the waters of Antarctica to Australia. This would enable engineers to test techniques of towing and harvesting, without having to deal with the problem of tropical waters, where little of the ice would be likely to survive.

There the idea rested as of the start of the 1980s. Still, the enticing promise of transforming the menace of icebergs into precious water is not likely to disappear. The difficulties are such, says Wilford Weeks, that "iceberg water advocates certainly have enough problems to keep themselves occupied for some time to come." But John Hult and other scientists are convinced that iceberg transport will become a reality before the 20th Century is out. Ω

Those who marvel at the immensity of icebergs sometimes liken them to impenetrable fortresses, to ancient cathedrals or mosques, even to the centuries-old Roman Colosseum. But such similes of strength and endurance are less accurate. From the moment they are released from the glaciers and ice shelves of the polar regions, icebergs drift helplessly toward doom. Despite their Gargantuan proportions, most icebergs disappear within two years, and few last longer than 10.

The death of an iceberg is one of nature's spectacles. Storm waves carve great bays and archways in the ice and slice deep channels with sheer cliffs on either side. Booming collisions of icebergs can cleave huge chunks from their flanks. All the while, the sun eats away the ice, and the atmosphere is alive with the sizzle and pop of compressed air bubbles, released from their icy prison after thousands of years.

As sun, wind and sea perform their destructive work, other forces operate deep inside the iceberg. Drifting the oceans' currents, icebergs shatter spontaneously into smaller pieces—a thunderous result of the presence of hidden fissures and weaknesses caused perhaps hundreds of years before, when the glaciers from which the icebergs calved twisted over the earth's surface on their way to the sea.

The iceberg's final throes begin with a great upheaval when the aging giant, sculpted and melted into fantastic grotesqueries, suddenly loses its equilibrium and flips bottom up. The toppled iceberg establishes a new center of gravity for a time. Then, melting into a new shape, it may lose its stability and capsize once more until, in a series of slow-motion somersaults, it melts away.

Three tabular icebergs, recently calved from Antarctica's Ronne Ice Shelf, inch their way into the Weddell Sea. Despite their imposing appearance, these icebergs—each one about a mile long—are midgets; icebergs more than

Its cliffs crumbling and its interior chiseled by the action of waves, a melting iceberg drifts off the coast of Antarctica. The horizontal bands indicate yearly accumulations of snow.

Its sides washed smooth by the sea, an
eroding iceberg rides the Antarctic Circumpolar
Current eastward around the continent.
Penguins perch briefly atop the ice before
continuing their search for fish.

The creation of this iceberg's startling profile began when the berg became grounded in shallows near Antarctica. The portion just above the water line has been eroded away by successive high tides.

Flipped over to expose a dazzling emerald underside, an aged iceberg lies stranded in the shallows off Livingston Island, near Antarctica's Palmer Peninsula. The green color comes from the iron and copper compounds of the fine rock debris trapped in the ice.

113

Chapter 4

PIONEERS IN ANTARCTICA

Great God! This is an awful place," wrote Robert Falcon Scott after he penetrated to the South Pole in 1912. The British explorer stated the case exactly. No environment on earth is so unrelentingly hostile to life as the heart of Antarctica.

The seventh continent consists of two gigantic domed ice sheets that are 15,000 feet thick in places and cover 5.5 million square miles. This desolate world of ice is larger than the United States, Mexico and Central America combined, and by all odds it is the coldest, the highest and the windiest of the world's great land masses. The mean monthly temperature at the South Pole in summer is −28° F.; in winter it is −80° F. In some areas the temperature has been known to drop to a brutal −127° F. Even during the summer months it rises above freezing only here and there on the Antarctic coastline.

Moreover, this is a dry cold—the driest imaginable. Even though 70 per cent of the earth's fresh water is locked up in the ice there, Antarctica is technically a desert, only slightly less desiccated than the Sahara or Gobi. The annual average snowfall of 25 inches is the equivalent of about four inches of rain; accumulation proceeds because there is virtually no melting from year to year, or from century to century.

The parched and bitter Antarctic landscape is made even more grim by barren mountain peaks that soar 17,000 feet above their collars of ice. And finally, terrible winds howl and shriek off the ice-laden mountains and plateaus at velocities of up to 200 miles per hour, stronger by far than those in the hurricanes and typhoons that churn through tropical seas.

In this world of superlative harshness, life is an anomaly. The vast herds of seals and teeming flocks of penguins and other birds that nest on the shores of Antarctica are actually creatures of the sea; they depend upon the ocean for sustenance and warmth. The glacial continent itself supports only an occasional insect, two species of primitive flowering plants, and a variety of mosses and lichens that cling to the rare ice-free outcroppings of rock, drawing their nourishment from bird guano.

How man came to know this fearsome place is an epic of perseverance in the face of nature's power. The valiant line of heroes who probed the mysteries of the glaciers suffered agonies of extreme cold, hunger and psychological disorientation. A number of them lost their lives. Yet the explorers and scientists continued to investigate this ultimate wilderness, armed with increasingly sophisticated instruments and survival gear, until at last they came to terms with the most forbidding of all glaciers.

Their hopes of being the first to gain the South Pole dashed, British Captain Robert F. Scott *(left)* and his weary companions stand before the evidence that Norwegian explorer Roald Amundsen had reached the Pole more than a month before them —on December 16, 1911.

Antarctica was discovered scarcely more than a century and a half ago, but its existence had been suspected long before that. As early as 500 B.C., Greek scholars, passionately concerned with symmetry, postulated a southern continent that would balance the known land masses of the north. They called it *Antarktikos*—the opposite of the cold region in the north, which lay under the constellation they called *Arktos,* or "Bear."

During the 18th Century, when explorers began actively searching for this southern continent, they were drawn by a number of often-conflicting desires. Their motives—and those of the pioneers who followed—were well summed up by the medieval Vikings who had roamed the Arctic waters at the other end of the earth. The lure of exploration, wrote the anonymous Viking authors of *The King's Mirror* in the 13th Century, "arises from the threefold nature of man. One part is a desire for contest and fame, for it is the nature of man to journey where there is hope of great danger and from it to obtain honor and praise. Another part is the desire for learning, for it is also man's nature to investigate those things he has been told, to see whether they are as he has been told or not. The third is the desire for wealth."

The first explorer to approach Antarctica was the British navigator James Cook, who sailed to the South Seas ostensibly with a commission from the British Admiralty to study astronomy from the vantage point of the Southern Hemisphere. But Cook was also under secret orders to search for the mysterious continent that was labeled on the maps of the day as *Terra australis nundum cognita*—"Southern land not yet known." (The term *"Antarktikos"* had faded when the medieval Christian Church supplanted Greek with Latin, and it would be revived only when the continent was discovered.) On January 30, 1774, Cook sailed his 462-ton sloop *Resolution* to within 300 miles of the Antarctic coast before heavy pack ice forced him to turn back. Cook could not see the continent that lay beyond, though he

On the theory that the spinning globe must have counterbalancing land masses, this 17th Century map shows two undiscovered continents at the North and South Poles. *Terra australis nundum cognita* (Southern land not yet known) turned out to have a basis in fact, but seekers of the northern *Terra septentrionalis incognita* (Unknown land under the seven stars) found only the frozen Arctic Ocean.

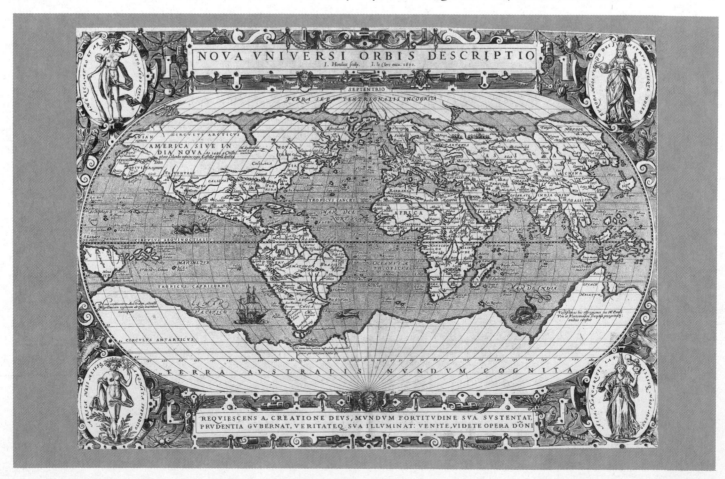

believed one was there. "The greater part of such a continent," he wrote, "must lie within the Polar Circile, where the sea is so pestered with Ice that the land is inaccessible. Thick foggs, snow storms, Intense Cold and every other thing that can render Navigation dangerous one has to encounter, and these difficulties are heightened by the Enexpressable horrid aspect of the Country, a Country doomed by Nature to lie buried under everlasting ice and snow." Glumly Cook concluded: "I will be bold to say that the world will not be benefited by it." The government took him at his word and put aside any further thought of territorial acquisition.

Cook was a marvelous sailor but a terrible prophet. Merchants—not only in Britain, but also in the newly formed United States, France and Russia—took note of his report that southern waters held an abundance of fur seals, prized for their oil as well as their pelts. By 1820 the seal population of northern waters had been virtually wiped out, and entrepreneurs began heading for the untapped riches at the other end of the globe. Some of them sighted Antarctica from time to time, but competition for hunting grounds was so fierce that the sealers kept their information to themselves. In any event, by the 1830s mariners were routinely plying the forbidding Antarctic waters, gaining confidence and pressing ever southward whenever the great ice pack permitted.

Governments were growing bolder as well. Between 1837 and 1841, France, the United States and Britain all sent out exploratory expeditions. In nearly simultaneous thrusts, the French and the Americans made the first official sightings of Antarctica by finding indentations where the ice pack extended only a few miles offshore. Beyond, stretching to the horizon, loomed the tall mountains of what was obviously a continental land mass.

The French leader, a Naval officer named Jules Sébastien César Dumont d'Urville, perceptively noted that the face of the land "appears to consist of a formidable layer of ice, rather like an envelope, which forms the crust over a base of rock"—a description quite satisfactory to modern glaciologists. Both d'Urville and the American commander, Lieutenant Charles Wilkes, charted stretches of coastline, and before retreating they proudly planted their national flags on rocky islets at the edge of the ice pack.

These expeditions were small, haphazard affairs compared with the venture that was mounted by the British Admiralty in 1839 and commanded by Sir James Clark Ross. His undertaking was not expressly intended for exploration; he was commissioned by the Admiralty to establish weather stations and study magnetic variation in the southern latitudes, and to find the South Magnetic Pole. But his orders were loose enough to allow him to use his own judgment, and Ross intended to do just that. Sealers had told him that on the Pacific side of Antarctica, at about the 180th meridian, a broad, lagoon-like expanse of open water lay beyond the ice pack. No ship had ever penetrated the pack, but Ross intended to ram his ship right through it. He hoped that the open water would lead to the South Magnetic Pole.

Of all the men yet to assault this unyielding end of the earth, Ross was the best prepared—in experience as well as equipment. The scion of a Navy family, he had first put to sea as a midshipman at the age of 12. Now, nearing 40, he was a veteran of 17 summers and eight winters in the Arctic, where he had participated in searches for the elusive Northwest Passage and had found the North Magnetic Pole. For his Antarctic expedition he chose two vessels that were perfect for the job. Both were round-bottomed Naval storeships designed to carry mortar shells and mines, and thus capacious enough to hold supplies for a long expedition and sturdy enough to withstand the ice. Ross named his two ships the *Erebus* and the *Terror*.

The expedition left Tasmania in November 1840. Within a few weeks, Ross was in the southern ocean, and on December 30 the southeast horizon showed a so-called blink—a shimmering band of light that told the voyagers the ice pack was not far off. Another five days' sailing brought them to its edge. Ross took his ships straight into it.

It was easy to see why other mariners had been scared away. "Our ships rolled and groaned amidst the blocks of ice, over which the ocean rolled its mountainous waves, throwing huge masses one upon another, then burying them deep beneath its foaming waters, the while dashing and grinding them together with fearful violence," Ross wrote. But the ships stood up to the punishment. By nightfall they had penetrated 50 miles through the ice, and in five days, after covering nearly 100 miles and reaching lat. 69° 15', they emerged into open water—the "lagoon-like expanse" of the sealers, kept open by swift currents and now known as the Ross Sea.

Ross and his men were rewarded with the sight of land on the horizon. "I could just trace the outline of a mountain, with a steep escarpment," wrote Robert McCormick, the flagship's doctor and the expedition artist. "It extended from southeast to southwest; very high, and was enveloped in snow. The whole of the upper part of this vast mountain range appeared to be a glaciation, relieved at intervals by the apex of some dark hummock or peak." McCormick proudly concluded: "We had discovered a new land of so extensive a coastline and attaining such altitude as to justify the appellation of a Great New Southern Continent." Ross named it Victoria Land for his Queen. The following morning he and a party of his officers and seamen planted the British flag on an offshore islet, which Ross called Possession Island.

Returning to their ship, the group pressed south, with the land three or four miles off the starboard bow. At about lat. 76° S., the men were treated to an astonishing sight: "a dense column of black smoke intermingled with flashes of red flame," as McCormick described it. A volcano was erupting in the icy fastness. Ross triangulated the height of the volcano—12,400 feet—and named it Mount Erebus in honor of his ship; he called a nearby peak of approximately 10,000 feet Mount Terror, for his companion vessel. The island on which the mountains stood was later to be named for the explorer himself.

After continuing southeast for about 60 miles, Ross came upon another surprise. His southbound course was suddenly halted by a mammoth white wall of ice. "It presented an extraordinary appearance," Ross recorded in his diary, "gradually increasing in height as we got nearer to it, and proving at length to be a perpendicular cliff of ice, between 140 and 200 feet above the level of the sea, perfectly flat at the top and without any fissures or promontories on its seaward face. What was beyond it we could not imagine, but it was an obstruction of such character as to leave no doubt upon my mind as to our future proceedings; for we might as well try to sail through the cliffs of Dover as penetrate such a mass." He was describing the phenomenon now known as the Ross Ice Shelf—the 500-mile-long skirt of floating glacial ice that extends into the sea from the Antarctic continent between the meridians of 170° E. and 158° W.

All this while Ross had been hoping for a shift in the coastline that would allow him to go west again—the direction in which he estimated the South Magnetic Pole to lie. Now he could go neither west nor farther south, so he turned east and sailed along the edge of the shelf. After two weeks his ships had traveled 250 miles, and still there was no opening that would allow them to resume their southward penetration. The shelf only rose higher, its end nowhere in sight. By now winter was closing in. Fearing that he might

A grizzly-bear pelt adds a rakish touch to the full-dress uniform of British Naval officer and polar explorer James Clark Ross in this 1834 portrait. Seven years later, he penetrated Antarctic pack ice to reach the open waters now called the Ross Sea.

be locked into the pack ice, Ross turned and retraced his path along the shelf. After a brief pause at a sound he named McMurdo for Archibald McMurdo, senior lieutenant aboard the *Terror*, Ross reluctantly gave up the quest for the magnetic pole and headed north, forcing his way back out through the ice and sailing to Tasmania.

Ross paid another visit to the ice shelf the following summer, but he could not find a chink in the wall that would take him south. Nonetheless, when he finally reached home in the fall of 1843, he was given a hero's welcome for having shown that the pack ice could be penetrated and that the mysterious glaciated continent was approachable.

Ross's epic voyages capped an era. Antarctica had been shown beyond question to be a continent, immense in size and smothered in ice, except where a volcano or two had managed to burn its way through the overlying glacier. Courageous men could no doubt land on it and examine its secrets at close hand. But the 19th Century was an age of empire, and the Antarctic appeared to offer few opportunities for colonization and exploitation. With the passing of the first wave of explorers, interest in the southern continent diminished for the remainder of the century.

In this contemporary engraving, the two ships of James Ross lie at the base of 180-foot cliffs—the edge of the immense ice shelf he discovered in 1841. A blowing whale and a seal in the foreground represent the marine life Ross found in Antarctic seas.

Though unexplored, the continent did not remain unvisited. Without exception, mariners in the environs of the Antarctic had noted an abundance of whales cavorting in the southern ocean—and whales were much in demand. Their oil greased the wheels of the Industrial Revolution; their blubber served as tallow, and their whalebone, or baleen, provided the stays of women's corsets, men's stiff collars and umbrella ribs. Like sealers, whalers eventually exhausted their hunting grounds in the north, and so they had to look elsewhere. Of 736 whaling vessels that New England ports sent out in 1846, most headed south.

The whalers ranged the entire circumference of Antarctica, often plowing through the pack ice into the Ross Sea and other open waters in search of their prey. From time to time, parties of men landed on the offshore islands, sometimes to vary their diet with the penguin eggs they found in abundance along the shores, and sometimes merely out of curiosity. In 1895 a whaling vessel dropped anchor just off Cape Adare on the continent itself, and several of its crewmen wandered ashore. One of them was Carsten Borchgrevink, a young Norwegian schoolteacher, who found—and recognized—a patch of pale green lichen growing on the rocks. It was the first evidence that Antarctica could support plant life.

Four years later, Borchgrevink, by now consumed with a passion for Antarctica, mounted his own private expedition to the frozen continent. He and his nine companions spent the entire winter in a hut on the beach at Cape Adare. One man died, and the others suffered unspeakable privations in temperatures that dropped to −50° F., winds that rose to 85 miles per hour and total darkness that lasted 72 days. In spite of the cold and the darkness, members of the party emerged from time to time to make invaluable meteorological and magnetic observations, to gather rocks and lichens and to find yet another form of life: three distinct types of insect.

What is more, before leaving Antarctica, Borchgrevink made a contribution of inestimable value to future exploration. Cruising along the Ross Ice Shelf, he came upon a spot where a gigantic iceberg had calved off into the sea, leaving a small bay four or five miles wide and a sloping face only 70 feet high. It was a relatively simple matter to scale the face—and to walk inland across the level plateau for about 10 miles to lat. 78° 50′ S., 45 miles closer to the Pole than anyone had ever ventured. The supposedly impenetrable fortress wall of the Ross Ice Shelf had proved to be not much of a barrier at all, and the shelf itself, extending far inland, would soon become the favored route of explorers heading for the Pole.

When news reached Europe that men had survived the rigors of winter on the Antarctic continent, an international mania for further probes was touched off. British, French, Swedish, German, Norwegian and Belgian adventurers vied with one another to achieve spectacular feats. To some extent they were motivated by competition for its own sake—a drive to assert both national prowess and personal heroism. But their zeal did not exclude scientific inquiry, and even the most ardent adventurers generally took along geologists, meteorologists, botanists, zoologists and physicists. Their findings, in turn, fed national pride.

Nevertheless, the overriding aim was to penetrate the Antarctic to lat. 90° S.—the South Pole. It came to represent the ultimate goal of exploration, and in quest of it a score of men risked—and sometimes lost—their lives in the first dozen years of the 20th Century.

The first expedition to follow Borchgrevink through the pack ice to the edge of the Ross Ice Shelf was mounted by the British, under the auspices of the country's most prestigious scientific organizations: the Royal Geographic Society and the Royal Society. One mandate of the mission was to conduct a reconnaissance of the continent's interior. The sum of £92,000 was raised—from wealthy patrons, from Parliament, and from the government of Australia. The Admiralty contributed men and matériel. A new ship, the 700-ton *Discovery,* was built especially for the expedition and equipped with powerful coal-fired auxiliary engines to augment her sails. Among the explorers she carried were five Naval officers and five civilian scientists—a botanist, a zoologist, a geologist, a physicist and a biologist. The *Discovery* also carried some novel equipment that would give the explorers an edge over their predecessors: a gas balloon to enable them to see from aloft terrain they could not ascend on foot, and a camera to record their findings.

The man selected to lead the *Discovery* expedition was Robert Falcon Scott, a 33-year-old Royal Navy lieutenant who was promoted to commander for the occasion. He had no previous experience in either of the polar regions and no established interest in them; the president of the Royal Geographic Society had, for reasons known only to himself, plucked Scott from the vast ranks of applicants. The *Discovery* was his first command.

In spite of his inexperience, Scott quickly showed himself to be a man of remarkable determination and physical stamina. An expedition member called him "the strongest combination of a strong mind in a strong body that I have ever known." Others of his men found him moody, introspective and given to holding grudges. As a Navy officer he was a bit of a martinet; indeed, a Royal Geographic Society scientist observed tartly that "Scott

The Elusive Magnetic Pole

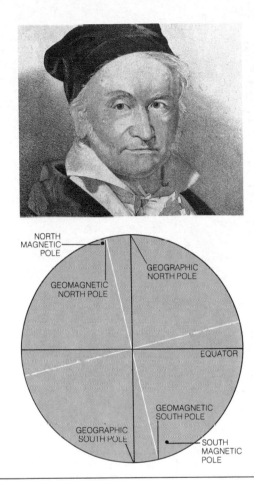

NORTH
MAGNETIC
POLE

GEOGRAPHIC
NORTH POLE

GEOMAGNETIC
NORTH POLE

EQUATOR

GEOMAGNETIC
SOUTH POLE

GEOGRAPHIC
SOUTH POLE

SOUTH
MAGNETIC
POLE

German mathematician Carl Friedrich Gauss *(right, top)* theorized that the geomagnetic poles would be located at opposite ends of an axis tilted 11½° from the earth's own axis of rotation. The actual magnetic poles, however, were found somewhat farther afield.

As early as the 15th Century, mariners realized that their north-seeking compass needles were not indicating the direction of the geographic North Pole but were pointing toward a magnetic pole, whose exact location was unknown. In 1831, British explorer James Clark Ross finally tracked down this navigation quarry: It lay 1,300 miles south of its geographic counterpart.

Around the same time, an eminent German mathematician, Carl Friedrich Gauss, attempted to predict the location of both the North and South Magnetic Poles. Using measurements of the earth's magnetic field, Gauss developed an equation that located a theoretical "Geomagnetic South Pole" *(diagram)*. But in 1909, members of Ernest Shackleton's British expedition discovered that the actual South Magnetic Pole lay almost 900 miles northeast of where Gauss had predicted.

Scientists now know that both magnetic poles are constantly moving—in response, many believe, to swirling electric currents within the earth. Since the poles wander independently, it is impossible to calculate the exact location of one from the location of the other.

considered ambition in a subordinate as little less than mutiny." For all these reasons, he provoked controversies that raged long after his time.

In February of 1902, Commander Scott sailed the *Discovery* into the Ross Sea and proceeded along the Victoria Land coast and then along the edge of the Ross Ice Shelf, pausing to land briefly here and there. He discovered that the shelf was more irregular than Borchgrevink had reported. In some places it had receded approximately 30 miles to the south, and in some it stood no more than 15 feet above the sea. At one stop, Scott climbed the slope with a couple of men and launched the gas balloon. Taking it up to 800 feet to survey the surface of the shelf, he saw a series of long undulations in the ice parallel to the shelf edge. Beyond them a white plain extended as far as the eye could see. One of Scott's officers took a turn aloft after him and photographed the scene.

After cruising the 500-mile length of the ice shelf, Scott decided to settle in for the winter and prepare for serious exploration in the spring. He put in at McMurdo Sound on the western edge of the shelf, in the shadow of the smoldering Mount Erebus, and erected a large wooden hut on a rocky promontory of Ross Island. With that as a base, Scott and his men spent most of the southern winter, from May to October, making short excursions to lay caches of heating fuel and food—biscuits, chocolate and pemmican, a mixture of dried beef and fat—along the route they would follow

when the push south began. The men were neophytes and made frequent mistakes at first. "Our ignorance was deplorable," wrote Scott. "We did not know how much or what proportions would be required as regards to the food, how to use our cookers, how to put up our tents, or even how to put on our clothes."

On November 2, Scott began his drive south. His companions were Dr. Edward Wilson, a zoologist, artist and the expedition's physician, and Reserve Lieutenant Ernest Shackleton, an energetic, outgoing Irishman of great personal magnetism. The three men took along 1,700 pounds of equipment on five sledges, with 19 dogs to drag the load.

Scarcely had the explorers set out across the Ross Ice Shelf when they encountered the first of a number of frightful problems. The animals they were depending upon seemed as weak as puppies. "After a few yards of struggling the dogs seemed to lose all heart," Scott wrote in dismay. So the men had to resort to relays—hauling three sledges for a few miles, then going back for the other two. The relay operation took four and a half hours to cover little more than three miles. Still the dogs were having great difficulty, so the men joined them in the sledge traces. "It is drag, drag, and drive, drive from the time we get up until it is time to turn in," Shackleton wrote. Part of the trouble, they decided long afterward, was the Norwegian stockfish they had brought from home to feed the dogs; it had spoiled while they were crossing the tropics and had been slowly poisoning the dogs ever since. Eventually they took to killing the weaker dogs and feeding them to the stronger. In the meantime, to ease the poor animals' burden, they cached some of their stores, leaving only a bare minimum of food on the sledges, and struggled on.

The men were sorely afflicted. The endless snow and ice reflect so much solar energy that they warm a layer of air whose uppermost edge causes the

Hut Point, named for its 36-foot-square bungalow, was home base from 1902 to 1904 for the first of British explorer Robert Scott's two expeditions to Antarctica. In the winter, Scott's party lived on their ship, the *Discovery*, which stayed anchored in McMurdo Sound.

light to refract and form mirages. Time and time again, Scott and his weary colleagues thought they had only a short march to a landmark that was in reality many miles distant. The glare of the infinite whiteness brought on bouts of snow blindness, a temporary but painful condition that was particularly hard on Wilson, who as the expedition artist never took his eyes off the scene.

The overcast skies sometimes brought a whiteout—a phenomenon that all the men came to dread. Sunlight diffusing through the clouds and reflecting back and forth between ground and clouds obscured the horizon and made everything seem pure white. The effect disoriented the men, made them dizzy and caused them to stumble on the uneven surface of the ice shelf. The wind cracked their skin and threatened frostbite. And they were forever hungry. Their labors required more food than they had brought, and after a while the shortness of rations began to affect their minds. Wilson wrote deliriously in his diary of "sirloins of beef, cauldrons full of steaming vegetables, but one spends all one's time shouting at waiters who won't bring one a plate of anything, or else one finds the beef is only ashes when one gets it."

Against all those punishments the men pressed on, and after 23 days Scott noted in his diary that they had reached a record 80° S. In the distance, some 60 miles to the southeast, the explorers could see a chain of mountains, and after another few days Wilson recorded: "It seemed to consist of a series of fine bold mountain ranges with splendid peaks—all snow clad to the base, of course, but here and there rocky precipices too steep to hold the snow stood out bold and dark. It was a wonderful sight, the pale blue shadows on the white ranges standing against a greenish sky." They were looking at the transantarctic mountain range, which bisects the continent and is now known to hold back the ice sheet that blankets the Pole.

By the 29th of December they had reached 82° 16′ 33″ S.—238 miles farther south than any of their predecessors had gone, and one third of the way to the Pole. But they were down to 10 dogs, which could no longer pull at all, and two weeks of food for themselves. For some time they had been feeling more and more exhausted, and Wilson, discovering they had swollen gums, realized they all had scurvy. Shackleton, who was suffering the worst of the three, was also short of breath and coughing blood. On New Year's Eve, a bitterly disappointed Scott decided that they could go no farther and must turn back.

The return was agonizing. The remaining dogs died off or were killed, and the men shed equipment until they were pulling only two sledges. After a time Shackleton grew too weak to share in the pulling; one morning he could not even walk and had to be carried on a sledge by the other two men—something that Scott would neither forget nor forgive. But miraculously, on February 3, after 13 weeks on the ice, they limped into camp, to loud cheers from the company they had left behind. "A lot of photographs were taken," Wilson wrote, "and indeed we must have been worth photographing. I began to realize then how filthy we were—long sooty hair, black greasy clothes, faces and noses all peeling and sore, lips all raw—everything either sunburnt or bleached."

In their absence, a relief ship had arrived from London with fresh supplies of coal, food and clothing. When the vessel was ready to depart, Scott ordered the ailing Shackleton to return home for medical care. Shackleton protested bitterly, but Scott commanded him to leave—and then perversely, when he wrote *The Voyage of the Discovery*, his memoir of the expedition, implied that Shackleton had quit for want of stamina. The implication was grossly unfair, as Shackleton would demonstrate in due course.

Ascending to 800 feet in a hydrogen-gas balloon, Robert Scott reconnoiters a possible route to the South Pole on February 4, 1902. The balloon soon "commenced to oscillate in a manner that was not at all pleasing," and Scott ordered himself hauled down.

Scott and the others stayed on for another year of less dramatic but tremendously valuable scientific study. The explorers surveyed the coast around McMurdo Sound, took soundings in the Ross Sea, and observed the habits of the whales, seals, penguins, skuas and fish that lived in the Ross Sea. One excursion along the coast of Victoria Land lengthened the list of life forms in Antarctica: A bloom of greenish algae was spotted growing in a pond near a mountain glacier. Another foray turned up an anomaly in the coastal mountains—an ice-free valley. It was "a valley of the dead," Scott wrote, where "the great glacier which once pushed through it has withered away."

The explorers' sense of excitement and surprise had diminished somewhat, however; when Scott wrote *The Voyage of the Discovery* he devoted 476 pages to the first year and only 190 to the second. On February 14, 1904, two relief vessels arrived with orders from the British Admiralty to bring the expedition to an end and return home. Scott did not protest.

In the meantime, Shackleton had been chafing under the humiliation of having been invalided home. The publication in October 1905 of *The Voyage of the Discovery* added insult to injury; Shackleton felt that the book made him a scapegoat for Scott's failure to reach the Pole. From that moment, he began raising money for an expedition of his own.

The task was not easy; Shackleton lacked the support that Scott derived from his position in the Royal Navy. But finally he won financial backing,

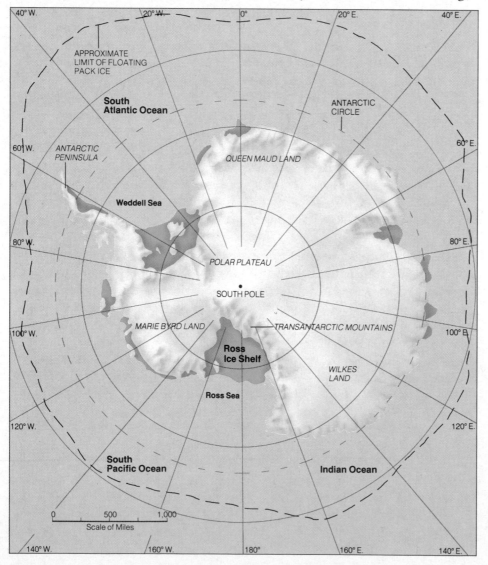

A map of Antarctica includes the continent's formidable girding of pack ice. British explorer James Clark Ross broke through this obstacle to reach the Ross Sea in 1841, and succeeding expeditions found that the Ross Ice Shelf offered the shortest and easiest route to the South Pole.

much of it from a wealthy industrialist, William Beardmore. Shackleton's plans were announced in *The Times* of London and in the *Geographical Journal.* Soon afterward he received a letter reflecting the depth of Scott's hard feeling: Shackleton's former commander forbade him to use the camp on Ross Island because he, Scott, was planning another expedition himself.

Shackleton shrugged it off and continued with his preparations. He bought a 40-year-old whaling ship, the *Nimrod,* and gathered a ship's company of 39 men. He recruited another 15 men to go ashore with him, planning to divide them in three groups: One would make an expedition east, another would head west and a third would go south, toward the Pole. Naturally, Shackleton would lead that one himself. For transport he took along nine dogs (though he had little faith in them after the experience of Scott's expedition), an automobile with a 15-horsepower gasoline engine and steel-ribbed tires, and 10 Manchurian ponies. "A pony drags as much as 18 dogs," Shackleton enthusiastically told an interviewer, "and consumes only 11 pounds of food per day as against 36 pounds of food required by 18 dogs." The logic was unassailable, but the reality would be bitterly disappointing.

Arriving at McMurdo Sound in late January, 1908, Shackleton made camp on Ross Island a few miles north of Scott's old base. He and his men built a sturdy hut, then settled in for a taxing winter. Only four of the ponies remained. One had not survived the voyage, another had been hurt in the landing and had to be destroyed, and the rest died from eating feldspar-crystal fragments and lava chips on the beach where they were tethered. The men tried to use the car to lay out the stores for the coming expeditions, but its wheels kept sinking in the snow. They decided that weight was the problem and stripped the vehicle down to its bare chassis and the driver's seat. In that form it succeeded in carting 750 pounds of supplies on a depot-laying mission or two, but it was clearly too troublesome to take on the long journey to the South Pole.

On October 29, 1908, Shackleton was ready to head for that goal. He took three companions—Lieutenant J. B. Adams, a meteorologist; Dr. Eric Marshall, a surgeon and cartographer; and Frank Wild, a factotum who had been on the Scott expedition six years before. Leaving the dogs behind at the camp, they hitched each of the four surviving Manchurian ponies to a sledge laden with 650 pounds of supplies. Shackleton estimated the distance to be 800 miles and expected the round trip to take 13 weeks.

The ponies proved unequal to the task. One went lame the first day out; though it limped along, it could do no work for a time. It had no sooner recovered and been reharnessed than the other three got so weak they had to be shot, and the men had to join in pulling the sledges by hand. Otherwise, the expedition proceeded well enough; in less than a month the explorers traveled 300 miles from their camp and passed the "farthest south" record established when Shackleton had accompanied Scott six years before. By December 6 they had reached the interior edge of the ice shelf and faced a 13,000-foot barrier, the Transantarctic Mountains, which Scott had seen from a distance. As the men searched for a pass, Shackleton wrote, "there burst upon our view an open road to the south, for there stretched before us a great glacier running between two mountain ranges." The glacier, which he named for his backer, William Beardmore, proved to be a 140-mile-long river of ice, one of eight large outlet glaciers that flow through the gaps in the mountains to provide a major proportion of the ice in the Ross Ice Shelf.

The Beardmore Glacier would provide access to the south—but not without a supreme test. It was lacerated with crevasses caused by the stress of its rapid downhill movement, and the last of the ponies disappeared into one of these chasms. Shackleton and his men had to pull the sledges by hand as they struggled up the steep glacier.

One evening when the exhausted explorers paused for dinner and a night's rest, Frank Wild made a discovery that would puzzle the scientific community for decades—and would ultimately help prove a bold and sweeping theory of the planet's mechanics. Near the head of the glacier, on an ice-free outcrop of sandstone, he noticed half a dozen eight-foot-thick seams that, unbelievably, looked like coal deposits. Coal, the fossilized remnant of lush green vegetation, seemed utterly out of place in this icy wasteland. But coal it was—as would be confirmed by scientists three years later—and its discovery was the most revolutionary of the decade-long race for the Pole. But it was left for later scientists to explain the puzzling

Using shaggy Manchurian ponies to pull their sledges, British explorers under Ernest Shackleton, a veteran of Scott's failed first expedition, renew the assault on the South Pole in November of 1908. This photograph was taken at the edge of the Ross Ice Shelf, more than 800 miles from the Pole.

implications of the find; Shackleton and Wild simply loaded some samples on a sledge and the next morning trudged on.

After three weeks of climbing, they reached an altitude of 10,000 feet and emerged onto a plateau that led without obstruction toward the Pole 300 miles beyond. They now had more or less level ground to walk on, but their ordeal was not over. Their food supply had run perilously low, and for the next three weeks they fought hunger as well as headaches and shortness of breath caused by the altitude. The weather turned foul: One day a howling blizzard held their advance to barely four miles. By January 9, 1909, they were at lat. 88° 23' S.—a tantalizing 112 miles short of the Pole. But to push on would have been suicidal. They might have made it to the Pole, but they would certainly not have made it back to their base. "We have shot our bolt," Shackleton wrote in his diary. He is said to have told his wife later: "I thought you would rather have a live donkey than a dead lion." He and his companions planted a British flag in the snow, took photographs and left a brass cylinder with a message to prove they had been there. Then they reluctantly turned around and headed north for the base camp.

The trek back took seven weeks and was itself a terrible test of endurance; the men were racked with hunger and dysentery most of the way. At one point they became so desperate for sustenance that they dug up the spilled frozen blood of one of the ponies they had shot and eaten on the outbound march, and concocted a broth from it. Bizarre as it was, Shackleton wrote in his diary that they "found it a welcome addition" to their pathetically meager rations.

They reached McMurdo Sound on February 28, 1909, after 117 days of struggle. Good news awaited them there. In their absence, their companions had not been idle; the team that Shackleton sent west had found the less glamorous but scientifically important South Magnetic Pole. They located it at 72° 25' S., 155° 16' E.—some 300 miles inland from the area where Ross had searched in vain for a break in the ice shelf. With its location discovered, navigation in the future would be more accurate.

To further brighten Shackleton's spirits, the *Nimrod* had arrived from New Zealand, where she had gone to winter, and was ready to take the

The 140-mile-long Beardmore Glacier gave Shackleton's party a passage through the forbidding Transantarctic Mountains and led them to a plateau 300 miles from the Pole.

explorers home. She set sail on March 4, 1909, to the accompaniment of three cheers and a rendition of "Auld Lang Syne" from her passengers as they left their base camp behind. Shackleton reached London on the 14th of June, 1909. If the public had ever shared Scott's ill feelings for him, such feelings were now dissipated; his countrymen gave him a hero's welcome, and the Queen conferred knighthood on the 35-year-old explorer for his glorious near-miss.

By this time Scott was already preparing another attempt to reach the South Pole. On his own return to London in September 1904 he had been promoted to captain, and with so much romance now surrounding polar exploration he easily raised funds from the government and private donors for a new expedition. For a vessel he selected the *Terra Nova*, a refitted whaler.

Scott left England in June 1910, thinking he had the field all to himself. But on his way south, when he stopped at Melbourne, Australia, to take on water and provisions, he received a cable indicating that the quest for the Pole was now a two-man race. The surprise entrant was Norwegian explorer Roald Amundsen. At 38, four years younger than Scott, Amundsen was a veteran of voyages to both polar regions. In 1897 he had gone south as second mate on the *Belgica,* a Belgian ship that had joined in the rush to Antarctica. The expedition had done some charting but otherwise contrib-

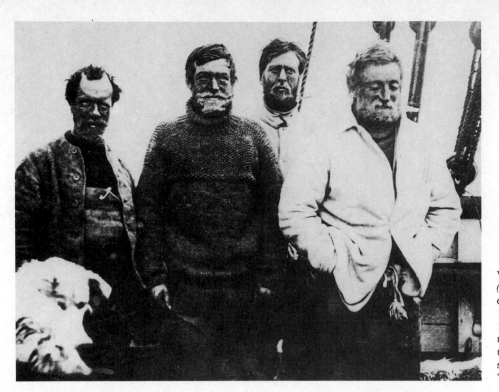

Weather-beaten and weary, Shackleton (*second from left*) and three other members of his expedition sail back to Britain in March 1909. The four had fought their way to within 112 miles of the Pole—417 miles closer than Scott's expedition six years earlier—before they were forced to turn back by dwindling supplies and by what Shackleton described as "blinding, shrieking blizzards."

uted no scientific discoveries of note. It had, however, made history of a sort by getting trapped in pack ice off the Antarctic Peninsula and unexpectedly spending a winter in Antarctic waters. Since then, the doughty Norwegian had become a highly successful commander in his own right. In 1905 he had fulfilled a centuries-old maritime dream by sailing a ship from the Atlantic to the Pacific through the ice-choked and island-studded waters atop North America; this conquest of the so-called Northwest Passage brought him world fame.

Scott was not the only one whom Amundsen surprised in the autumn of 1910; everyone else was surprised as well—including his own government. Amundsen's original intention was to be the first man at the North Pole. But in September of 1909, as he was organizing his expedition, he was informed that the American adventurer Robert Peary had reached the Pole on April 6 of that year. Amundsen, a man with a ferocious drive to win at whatever he did, simply switched objectives and decided to head for the South Pole. He craftily kept his change of plans a secret until he was well on his way to the Antarctic because he feared that the Norwegian government, intent on cultivating British good will, would forbid him to challenge the English hero.

The two expeditions and their leaders were a study in contrasts. Scott was short in stature; Amundsen was more than six feet tall. Amundsen was precise, dogmatic and single-minded; Scott was moody and emotional. Scott represented a rich and mighty empire, Amundsen a small, poor nation that had been a sovereign state scarcely four years (until 1906, Norway had been ruled by Sweden). In contrast to Scott's complement of 33 men of various backgrounds, Amundsen's crew consisted of only nine men, all specialists whose physical stamina and polar knowledge he could count on. One of Amundsen's men was a ski champion; another was an expert dog handler. His objective was simply to get to the Pole, and he did not confuse the issue with science.

Amundsen's determination was exemplified by the dogs he took along and the use he intended to make of them. He purchased 97 Greenland Huskies, and he trained them rigorously; an Englishman who observed

The active volcano of Mount Erebus forms a backdrop for British Captain Robert Falcon Scott in a 1911 photograph that was taken shortly before he set out on his race to beat Roald Amundsen to the South Pole.

Norwegian explorer Roald Amundsen radiates his iron resolve in a photograph taken on board a ship in 1903, two years before he proved the navigability of the fabled Northwest Passage and seven years before he headed for Antarctica.

them exercising noted that on the sound of Amundsen's whistle an entire team would stop as one dog. Amundsen considered the animals "quick, strong, sure-footed, intelligent and able to negotiate any terrain that man himself can traverse." But he did not allow an ounce of sentimentality to cloud his approach to them. He intended that in Antarctica the dogs should eat fresh-killed seals and penguins, and that if those provisions ran out the dogs should eat each other—and that, in dire circumstances, the men should eat the dogs.

Scott took quite a different attitude. After his first sorry experience with dogs, he would not rely on them again. To be on the safe side he took along a pack of 34 Siberian Huskies, but he paid no particular attention to their training. He put more faith in two other forms of power. One was a herd of 19 Siberian ponies—notwithstanding Shackleton's cautionary experiment. The second, viewed as most promising, was a trio of mechanized sledges with belted and cleated tracks designed to haul two-ton loads at 3.5 miles per hour over ice and snow.

Ultimately, however, Scott intended to depend neither on dogs nor ponies nor motor sledges, but on himself and his men. He held the sentimental view that it was preferable for men to get to the Pole on their own two feet. "In my mind," he wrote, "no journey ever made with dogs can approach the height of that fine conception which is realized when a party of men go forth to face hardships, dangers and difficulties with their own unaided efforts, and by days and weeks of hard physical labor succeed in solving some problem of the great unknown. Surely in this case the conquest is more nobly and splendidly won." That notion was to cost him dear.

In January 1911, these two disparate men arrived in the Ross Sea about a week apart and took up winter quarters at opposite ends of the ice shelf. Scott returned to Ross Island and, finding his old quarters made inaccessible by the ice that had accumulated in the harbor, settled in an area 15 miles north of there. Amundsen put in more than 400 miles to the east, and set up camp on the ice shelf where it terminated in the Bay of Whales, an inlet that Shackleton had found in 1908. It was a risky place to camp; if the shelf had calved, his base could have floated out into the ocean or been obliterated altogether. But Amundsen chose the site because it put him 60 miles closer to the Pole than Scott on Ross Island. For men traveling in Antarctica, 60 miles could be crucial.

At their separate bases, the Norwegians and the British went about the now-established routines of building living quarters and sledging out to place caches of supplies. Amundsen laid three tons of food in several depots spaced out over a distance of 250 miles, whereas Scott stored only one ton at a single depot 150 miles south of his base. Both men planned to place additional depots en route to the Pole.

At Amundsen's base, as the snow began to pile up in drifts around the hut, the men had a bright idea. Instead of shoveling the snow away, they burrowed into the drifts and carved out a warren of tunnels and snow caves that no wind could whistle through. As the winter months passed, the Norwegians busied themselves with improving their gear. They rebuilt the hickory sledges they had brought along, cutting the average weight from 160 pounds to 50. Using ink, they darkened the canvas they would use for tents along the march to reduce glare and absorb the rays of the summer sun. On these and other chores they worked six days a week, stopping only on Saturday night to sprint naked down the tunnel of one of the ice caves, where a sauna had been improvised with two primus stoves providing the heat. They celebrated the Sabbath with a ration of aquavit.

Not surprisingly, Amundsen was first off the mark when winter ended. On October 20, with the temperature at about 0° F., he started for the Pole, taking with him four companions, four sledges and 52 dogs. His main worry—and it gave him no rest—was the threat of Scott's three motor sledges; if the machines worked, even the best-handled dogs in the world might not beat them to the Pole.

Amundsen need not have fretted. One of Scott's motor sledges had plunged through the sea ice to the bottom of McMurdo Sound as it was being hauled ashore, and the other two were to break down during the first leg of the journey. At that, Amundsen enjoyed an almost insurmountable lead from the outset. Scott and his party of 16 men—including support teams with eight pony-drawn sledges to haul equipment and food along part of the route—did not get under way until the 1st of November, 12 days later than Amundsen. By that time Amundsen was already 230 miles nearer the South Pole.

During the next two weeks, Amundsen lengthened that lead. His dogs sprinted ahead while Scott's ponies sank up to their hocks in the snow; his men skimmed over the ice on cross-country skis that could bridge small crevasses while Scott and his men trudged step by weary step. "We are going like greyhounds," Amundsen exulted in his diary on November 8. He was averaging 20 miles a day, with plenty of time for rest, while Scott made only two thirds that speed with twice the effort.

On November 17 Amundsen reached the halfway mark. He came to the end of the ice shelf and stood at the foot of the Transantarctic Mountains at a point about 200 miles east of the Beardmore Glacier, Shackleton's avenue through the mountains. He found another glacial valley piercing the mountains and called it the Axel Heiberg Glacier, after one of his patrons. The glacier was shorter than the Beardmore by perhaps 100 miles, but climbing it proved to be a nightmare, for it was a steep cataract of ice, part of which rose 8,000 feet in only 20 miles. It was laced with crevasses and littered with great blocks of ice that had fallen off and tumbled down from above. The struggle up the perilous slope took Amundsen four days, but at last he reached the top and started across the rugged polar plateau.

Almost immediately the weather broke, and blizzards swept the plateau with punishing winds for more than two weeks. During one four-day period the men could only huddle in the lee of an ice block and pray for survival. When the wind finally dropped and the weather turned fair, the

A member of Roald Amundsen's 1911 expedition to the South Pole plants a black flag atop a snow cairn marking one of the supply depots placed at five-mile intervals along the way. The party cached provisions for the return journey so lavishly that the satiated explorers left a ton of uneaten food behind them.

Norwegians had clear sailing across some 100 miles of hard-crusted snow.

The climactic approach to the Pole was tense with expectation. On the night of December 13 the men camped at 89° 45' S., and calculated that they were only about 17 miles from the Pole. "It was like the eve of some great festival that night in the tent," Amundsen wrote. They took out the Norwegian flag and lashed it to two ski poles. "Then it was rolled up and laid aside to be ready when the time came."

Next day they set out again. A sighting of the sun at noon showed them to be at 89° 53'. Then, early in the afternoon, the lead sledge driver shouted to Amundsen, "Will you go ahead?"

"What for?" asked Amundsen.

"The dogs run better with someone in front of them," answered the man with sly psychology. It was a mark of their affection for Amundsen that his companions wanted him to be the first man literally to set foot on the South Pole. Amundsen took the lead, and shortly after 3 p.m., when their sledge meters told them they had covered 17 miles since the night before, a spontaneous "Halt!" arose from the men.

They half expected to see a British Union Jack on the site. But there was nothing—no feature to distinguish this long-sought locus from the endless whiteness that surrounded it.

Without a word, the five men shook hands with one another. Then Amundsen got out the Norwegian flag on its ski poles, and with each of the men gripping this improvised flagstaff, "five weather-beaten frost-bitten fists," as Amundsen described the tableau, planted it at the Pole.

The next day, Amundsen measured the sun's altitude with a sextant every hour from six in the morning till seven at night, and found that the ritual of the day before had been premature: They were still five and a half miles from the Pole. And so the following morning Amundsen and his men moved the camp and their flagpole, took another round of sightings and satisfied themselves that this time they were at exactly 90° S. Before leaving, they pitched a tent and affixed the Norwegian flag to its top. As a precaution against any misadventure on his return, Amundsen placed inside the tent a letter to King Haakon of Norway reporting his conquest, together with a covering letter to Scott asking him to forward the letter to the King. Then the explorers headed back to their base. They reached the coast on the 26th of January, 1912, having completed the epic 1,600-mile round trip in 98 days.

Meanwhile, Scott was in desperate straits. When the Norwegians stood at the Pole on December 16, the English expedition was still struggling up the Beardmore Glacier almost 400 miles away. As before, the ponies had fared so poorly that all eight had either died or been shot. Scott and his men were hauling the sledges themselves, pulling in pairs with the ropes hitched to canvas bands worn across their stomachs. The burden of pulling some 200 pounds per man was "simply jerking our insides out," wrote one of the members of the team.

On New Year's Day, 1912, Scott and his entourage emerged at last from the Beardmore Glacier onto the polar plateau, and their spirits rose; they did not know that, scarcely 100 miles to the east, the victorious Amundsen was on his way home. Three days later, Scott sent the last of the support teams back to the base camp and slogged on with four companions. They were 37-year-old Petty Officer Edgar Evans of the Royal Navy; Lawrence E. G. Oates, a 31-year-old cavalry captain; Henry R. Bowers, 29, a lieutenant in the Royal Indian Marine; and Scott's long-time companion, Dr. Wilson from his *Discovery* days.

What happened thereafter is one of the most tragic stories in the annals of polar exploration, made the more poignant because much of it can be read in Scott's own diary. The British reached the Pole on January 18, 1912, only to find the Norwegian flag fluttering atop Amundsen's tent. Scott was crushed—and bitterly affronted by Amundsen's request that he forward the letter to King Haakon. Amundsen, of course, had intended no insult; he was simply taking the precaution in case he himself failed to return. But Scott did not see it that way. A member of the expedition who did not go to the Pole later summed up Scott's plight by saying that the Amundsen letter degraded Scott "from explorer to postman."

Broken in spirit, short of food, suffering from frostbite and perhaps from scurvy, Scott and his men began the 850-mile journey back, dragging their sledges behind them. They faced two months of hideous torment. Without dogs to pick up the scent of their trail in, they frequently lost it and wasted valuable hours finding it again.

The first to die was Edgar Evans, who had been injured in a fall; he succumbed on February 17, at the foot of the Beardmore Glacier. A month later, Lawrence Oates, suffering grievously from frostbitten feet that had turned gangrenous, struggled from the tent one morning and deliberately walked away to a lonely death on the ice rather than remain a burden to the other men. Then on March 20, 1912, with winter fast approaching, Scott, Wilson and Bowers were stopped by a blizzard less than 13 miles from a supply depot stocked with food and fuel for their primus stoves. "Every day we have been ready to start for our depot," Scott wrote in his diary, "but outside the door of the tent it remains a scene of whirling drift. I do not think we can hope for any better things now."

The three men lay in their sleeping bags inside the tent for at least nine days while their lives ebbed away. With the last of his strength, Scott kept up his diary and wrote letter after letter—to his wife, whom he sorrowfully addressed as "my widow"; to old friends, even a "Message to the Public." "I do not think human beings ever came through such a month as we have come through," he said, "but for my own sake I do not regret this journey." To Sir James Barrie, the playwright who created Peter Pan, Scott wrote: "We are showing that Englishmen can still die with a bold spirit, fighting it out to the end." The last diary entry was dated March 29. "It seems a pity," Scott lamented, "but I do not think I can write more."

Eight months later a search party found the tent, with snow drifted over it and the three bodies inside, Scott's diary and letters beside him.

Ironically, Scott's failure stirred the world more than Amundsen's success. This was true not only because of the tragic circumstances in which Scott and his companions perished but also because of Scott's eloquence in recounting his cruel ordeal. His diaries and letters helped create a legend of nobility and heroic suffering.

As for the winner of the race, Amundsen wrote a matter-of-fact account that minimized the hardships of travel in the glacial wasteland. That is not to say he failed to savor his triumph. When some mean-minded critics suggested that he had been lucky while Scott had been a victim of misfortune, Amundsen replied: "Victory awaits those who have everything in order—people call this luck. Defeat awaits those who fail to take the necessary precautions—this is known as bad luck."

Amundsen's carefully planned life had a sad finale: In the following decade he died in an airplane crash in the Arctic Ocean. Romanticists find a poignant symbolism in the entombment of the old rivals—Amundsen and Scott—at opposite ends of the earth.

With the passing of Amundsen and Scott, the quest to unveil the mysteries of the glacial continent entered the era of modern technology, and the men who ventured onto the ice did so in large, magnificently equipped teams. In 1928, less than six months after Amundsen's death, Richard E. Byrd, a 40-year-old U.S. Naval aviator, set out for Antarctica. With him went 83 men on board two ships whose cargo holds contained 600 tons of supplies plus a snowmobile and the parts for assembling three aircraft, which Byrd intended to use for reconnaissance and aerial mapping of the immense continent.

Arriving at the Ross Ice Shelf in the last week of December 1928, Byrd proceeded to erect a veritable village that included three main structures, a dozen huts and hollowed-out snow caves to house his ski-equipped planes, his food and other gear. Little America, as Byrd named his headquarters, boasted such amenities as telephone connections throughout the camp, a motion-picture projector and some Charlie Chaplin films, and 50 gallons of alcoholic beverages (forbidden back home by the law of Prohibition).

Little America also boasted three 65-foot steel towers that marked the first successful use of radio in Antarctica. The radio enabled the exploration

Less than 175 miles from the South Pole, Captain Scott (far right) and three of his party pull a heavily laden sledge across the polar plateau. They dragged their own supplies—approximately 200 pounds per man—a total of 400 miles to their goal.

teams to keep in touch, coordinate their tasks, and call for help if they needed it. In addition, it brought the men welcome voices from home; Byrd had arranged that every Saturday afternoon at 4 o'clock they could receive programs specially beamed across 10,000 miles by powerful transmitters in the United States.

Before winter set in, the expedition tested the airplanes and learned some of the advantages and hazards of flying in Antarctica. Making a five-hour flight that took him 250 miles to the east of his base at Little America, Byrd discovered a new mountain chain. "Here was something to put on the map: a fine new laboratory for geological research," he wrote. But six weeks later, when three members of Byrd's team landed on the ice at the edge of these mountains, they found that aircraft were particularly vulnerable to the elements in this ferocious land. A sudden gale with 150-mile-an-hour winds swept down from the mountains, picked up the two-and-a-half-ton plane—guy ropes, chocks and all—and dashed it against the side of a mountain half a mile away. With the plane went the radio, and the men were stranded. They had to subsist on emergency rations for nine days before a rescue flight spotted the wreckage and landed to pick them up.

Despite his perilous initiation to flight in this glaciated land, Byrd made ready for the principal adventure of his expedition—an aerial survey of the route Amundsen had taken to the Pole 18 years before. Byrd took off on November 28 with a three-man crew consisting of a pilot, a radio operator and a photographer; Byrd acted as navigator.

The flight was fraught with danger. There was no telling how the plane might react aloft to a sudden storm, what the visibility might be en route, what sort of terrain they might encounter in a forced landing; and the aircraft radios were notoriously quirky. Taking precautions against the worst, Byrd wrote farewell letters to his family, friends and sponsors, and left sealed instructions to his next-in-command to be opened if he did not

The Transantarctic Mountains loom behind the Scott expedition's desolate campsite on the Beardmore Glacier, halfway to the Pole. Scott and his party hung their wet clothing from their skis to dry during an overnight stop before resuming their 1,700-mile trek.

Haggard with exhaustion, Scott (center) and his men line up in front of "our slighted Union Jack"—the second flag to fly at the South Pole. Lieutenant Henry Bowers (seated, left) took the photo by tugging a string fixed to the camera; the film was found with the men's belongings.

return. He also loaded the plane with survival gear and 250 pounds of food to last him and his crew for a month in case they were forced down.

And indeed, they encountered terrible danger almost immediately. The narrow Axel Heiberg Glacier, in the pass Amundsen followed through the Transantarctic Mountains, was shrouded with fog, so Byrd decided to try another route. The alternative, the Liv Glacier some 60 miles to the west, was clear, but it appeared to be at least 10,000 feet high—and with all the gear aboard, the plane was having difficulty climbing to 9,600 feet and threatened to stall. The pilot barked out orders to lighten the load. The photographer left his camera long enough to jettison a bag of food. The plane climbed, but not enough, so they dumped a second bag of food—the last of their survival rations. The desperate gamble worked. "The plane literally rose with a jump," Byrd wrote later. "We would clear the pass with about 500 feet to spare." Ahead lay the level expanse of the plateau, an open pathway to the Pole.

The plane flew on to the Pole and crossed it twice from different directions while Byrd confirmed the location with a sextant and dropped a weighted American flag as near as possible to the Pole. He and his crew then returned to Little America without further incident. The trip that had taken Amundsen 98 days took Byrd a mere 18 hours and 36 minutes.

The expedition remained in Antarctica for 14 months. By the time Byrd went home to a hero's welcome in mid-1930, he and his men had flown over an estimated 200,000 square miles never before glimpsed by man, and had photographed much of this by aerial camera—"which sees everything and forgets nothing," Byrd wrote. His principal discovery was the tract he named for his wife, Marie Byrd Land, which reaches from the eastern edge of the Ross Ice Shelf to the base of the Antarctic Peninsula.

Four years later Byrd returned to Antarctica with an even larger expedition. This time the team included 115 men representing 22 scientific disciplines, among them the new one of cosmic-ray research. They had six tractors, one of which could pull a load of 20,000 pounds. They also brought four planes and a newfangled creation called an autogiro, a precursor of the helicopter. The autogiro eventually crashed from an accumulation of ice on the tail, but not before making a number of useful short-range reconnaissance flights. So much public excitement had been generated by the expedition that a leading broadcasting company agreed to handle radio traffic to and from Antarctica free of charge. The magic of radio sent back to American homes an authentic sense of high adventure at the far end of the world, for Byrd had shrewdly arranged to make live weekly broadcasts from Antarctica to tell of his activities.

On this trip to Antarctica, Byrd conducted a daring experiment designed to test the human psyche—with himself as the guinea pig. He settled in a 9-by-13-foot cabin dug into the Ross Ice Shelf, 125 miles inland from the base at Little America, intending to spend six months absolutely alone. He would busy himself monitoring weather conditions, but his main objective was to see how he would adjust to this self-imposed solitary confinement—"to taste," as he put it, "peace and quiet and solitude."

Byrd took up his isolation on March 28, 1934, the beginning of a period of 24-hour darkness and temperatures as low as −80° F. He established a routine of making outdoor forays at regular intervals to check a battery of instruments for temperature, wind, precipitation, snow drift and mist, and to note cloud conditions. When he was not busy with these chores, he cooked meals, kept house, read, listened to records and marked off the days on a wall calendar. Three times a week he made radio contact with Little America. For about six weeks, all went well. Then he began to suffer severe

In the library at Little America in 1929, Commander Richard E. Byrd ties an American flag to a stone from the grave of pilot Floyd Bennett, who died shortly before he was to accompany Byrd to Antarctica. Byrd dropped the commemorative stone from his airplane as he flew over the South Pole.

headaches and pain in his eyes. He found that the fresh-air ventilator in the roof was clogged with ice and snow and that fumes were leaking through the pipe leading from his oil stove, poisoning him with carbon monoxide. Byrd made repairs, but a couple of weeks later he received another dose of carbon-monoxide poisoning from a faulty gasoline generator. He repaired that too, but still he was sick. He hurt all over and could not keep food down. He went into a steady decline, and by mid-July was on the verge of collapse. He began to think he was going to die. But he tried to conceal his plight from his colleagues at Little America; he feared that no rescue party could survive a trip to the cabin in the cold and constant darkness.

Back at Little America, Byrd's men noticed that his radio messages were becoming garbled; they knew something was dreadfully wrong and that they must go after him. During the first week of August a team of three set out on a tractor, and on August 10 they reached Byrd's lonely outpost. They found the cabin littered with rubbish and spoiling food, and Byrd haggard and wild-eyed. They had to nurse him back to health for more than two months before they dared subject him to the rigors of the return trip to Little America, and thence home to the United States. Doctors suggested that to the carbon-monoxide poisoning had been added acute depression. The phenomenon, unexpected in Byrd's time, was to plague future visitors to the Antarctic again and again.

When Byrd returned from Antarctica in 1935, he brought the heroic age of polar exploration to an end. He had privately financed both trips through prodigious promotional efforts; he had enlisted 100 volunteers to procure free supplies of meat, fuel, boots, tobacco, cereal, rope and twine; he had recruited and directed the teams that went with him; he had put his own distinctive stamp on the expeditions by such means as his radio broadcasts and the publication of a book describing his solitary ordeal. Such personal touches would soon vanish into history, along with the heroics of his predecessors Ross, Shackleton, Scott and Amundsen.

If he marked the close of one era, however, Byrd also foreshadowed the coming of another. In pioneering the use of radios and aircraft, he had shown what technology could do to advance man's knowledge and ease his way. And when polar exploration resumed after World War II, science came to Antarctica in earnest. Ω

CAPTAIN SCOTT'S GIFTED CHRONICLER

At the age of 39, Herbert G. Ponting already had 20 years of experience in photography when he met Captain Robert Falcon Scott for the first time in London in 1909. Charmed by the determined and persuasive explorer, Ponting signed on as the official photographer of Scott's expedition to the South Pole.

One year later the "camera artist," as Ponting described himself, stepped ashore on Antarctica. For the next 15 months he meticulously chronicled on film the triumphs and privations of the Scott expedition.

Nicknamed Ponko by the other members of the 33-man expedition, Ponting searched for magnificent images of the polar wilderness. "This world is a different one to him than it is to the rest of us," Scott wrote in his diary. "He gauges it by its picturesqueness."

Though his colleagues were constantly busy with scientific experiments and preparations for the dash to the Pole, Ponting often conscripted them to model. "To 'pont,'" joked one scientist, meant "to pose, until nearly frozen, in all sorts of uncomfortable positions."

Ponting was fascinated by the strange Antarctic fauna, and cheerfully braved hardships and danger to find and photograph seals, penguins and other birds. But he soon discovered that such work required patience and ingenuity. Because oil froze in the extreme cold, he used graphite to lubricate his equipment. If he touched his cameras barehanded, he risked instant frostbite.

Ponting stayed behind when Scott's party set out for the South Pole in November, and he left Antarctica four months later without knowing if they had succeeded. Concern for these men haunted him for months. After learning of their deaths, he at least had the comfort of knowing that he had created a lasting record of the heroic expedition.

Upon Ponting's death in 1935, one survivor of the expedition summed up his achievement. "He came to do a job, did it and did it well. Here in these pictures is beauty linked to tragedy— one of the great tragedies—and the beauty is inconceivable for it is endless and runs to eternity."

Near Mount Erebus on the shore of McMurdo Sound, the base camp of the Scott expedition defies a barren landscape in this picture taken by Herbert Ponting *(inset)* in March 1911.

Surrounded by decorative flags and dog-sled banners, Captain Scott *(center)* and his team celebrate Midwinter's Day—the 22nd of June—in the wardroom at the base camp. After relishing such delicacies as hot mince pie, chocolate and even a flaming plum pudding, the men broke out champagne to toast the halfway point of the polar night.

To pass the time at the end of a long, dark day, Ponting delivers a lecture on Japan and the Far East, showing lantern slides he took during his visits. "I was glad to find that these slides were much appreciated," he wrote, "and I believe getting back into the world again, for an occasional hour or two, had a healthy effect."

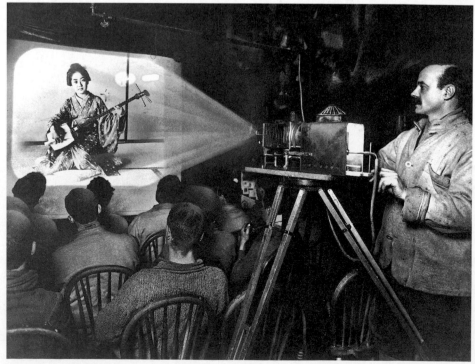

Thomas Clissold, the expedition's cook, rolls out crust for a special treat—rhubarb pie. The men enjoyed seal meat and fish, along with supplies of New Zealand mutton and a variety of canned goods. But they could not bring themselves to consume the enormous quantity of pickles that had been included in the stores. "One's taste does not run overmuch to pickles in such latitudes," noted Ponting.

Comfortably sheltered from the lethal polar night, five bunkmates enjoy a relaxed moment before turning in at the base camp. Other men made modest attempts to create homey sleeping quarters, but this billet, said Ponting, "was conspicuous by its lack of any attempt at anything more than necessary comfort, and was christened 'The Tenements.'"

In the small corner of the base-camp cabin designated for his laboratory, Dr. Edward Atkinson studies a test tube containing tiny Antarctic parasites. Atkinson, the expedition's surgeon, specialized in parasitology and the study of bacteria, but he was often in demand to treat a variety of ailments—especially frostbite, a condition he experienced firsthand when he was caught in a severe blizzard.

Canadian physicist Charles Wright shoulders an ice pick on a mild day in November. Wright conducted exhaustive studies of Antarctic ice, and he later wrote an extensive treatise on glaciers. Under Ponting's tutelage, Wright learned the techniques of photographing ice-crystal formations in the field.

Edward Wilson, chief of the 12-man scientific team, paints a watercolor of the moon's halo, caused by refraction in high-altitude ice crystals. Wilson sketched outdoors, then painted indoors. "Because of the nature of our work," wrote Ponting, "I was drawn into closer contact with Wilson than with any other of my comrades."

Biologist Edward Nelson adjusts the tripod he used to hoist up water samples and marine life from beneath the ice. Nelson found that the water temperature was a constant 29° F., regardless of the thickness of the ice or the air temperature. "Here he spent much of his time, weather permitting," Ponting said, "repeating each day unpleasant, wet and messy operations."

Bundled up to ward off the numbing cold, Royal Navy surveyor Edward Evans observes an eclipse of Jupiter through a telescope. Since the sun did not shine for four months, Ponting relied on magnesium flash powder for night photography.

George Simpson, the expedition meteorologist, records readings at the weather station he established on Wind Vane Hill near the base camp. Simpson's preoccupation with his many instruments led Ponting to call him "the wizard of our little community."

Crossing the Barne Glacier on Ross Island, three members of the expedition pause to investigate an ice bridge spanning a deep crevasse. "Covered crevasses can usually be distinguished easily enough when the sun is shining," Ponting wrote, "but it is vastly more difficult to detect them in cloudy weather."

On a photographic outing in McMurdo Sound, expedition leader Robert Scott pauses beside a ridge of sea ice. These pressure ridges took shape when the frozen sea was compressed and buckled by the forward motion of the Ross Ice Shelf.

Driven by fierce winds, frozen salt spray from McMurdo Sound created what Ponting described as "deep irregular furrows, as though the ice had been turned over by a gigantic plough."

Sculpted into the shape of a Norman keep, an iceberg dubbed Castle Berg gleams in the sun as expedition members approach. It was, Ponting decided, "a fitting palace for King Jack Frost, whose home I never doubted this to be."

"From outside, the interior appeared quite white and colourless, but, once inside, it was a lovely symphony of blue and green," Ponting said of this huge grotto that had been carved in a decaying iceberg by wind and waves.

Making new acquaintances, Ponting stands amid an Adélie penguin rookery seven miles from base camp. Ponting was enchanted by these birds and called them "the comedians of the south," but he knew they could deliver a nasty peck when offended. "It is well to get in a few bows at this time," he wrote, "and to affect the air of an Emperor, muttering after the manner of their kind as you do so."

Basking contentedly, a colony of seals sprawls on pieces of pancake ice in McMurdo Sound. Pancake ice forms when sea water, driven by frigid winds, becomes slush. When the wind subsides, the slush congeals into solid floes that then collide, resulting in a raised lip around the edges. Ponting observed many seals sun-bathing on the ice: "Often they seemed to be dreaming, for they would start in their sleep, and snort and gnash their teeth, whilst a quiver ran all over their sleek, floppy forms."

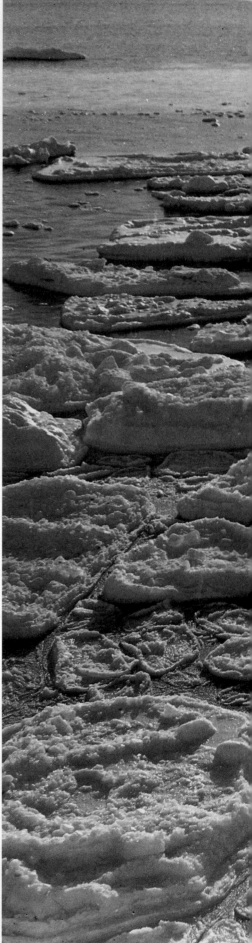

Chapter 5

A CONTINENTAL LABORATORY

Early in the morning of July 1, 1957, the dark horizon of the Antarctic ice sheet glowed blood red, and vast, shimmering yellowish green rays swept silently upward into the vault of the sky. A severe magnetic storm raging in the atmosphere was producing one of nature's most stupendous displays, the southern aurora. By richly symbolic happenstance, this particular extravaganza of the heavens did not play itself out for a few penguins in a virtually empty wasteland, but entranced hundreds of scientists scattered throughout the frigid wilderness. They were assembled in Antarctica for the International Geophysical Year, an 18-month program of intensive research into just such occurrences as the aurora—whose onset coincided precisely with the official beginning of their enterprise.

The International Geophysical Year, or IGY, was a concerted effort to make worldwide, detailed observations of such phenomena as terrestrial magnétism, cosmic radiation and the weather. The Poles, where the earth's magnetic field is most intense and its interaction with the atmosphere is most frequent and pronounced, offered the best vantage point for many of these observations. And the South Pole had the additional virtue of providing plenty of solid land from which to watch. Thus, while the IGY was global in scope, the major effort was made in the glacial reaches of Antarctica, by nearly 1,000 men from 12 nations working at more than three dozen stations throughout the continent.

The period from July 1957 through December 1958 was chosen for the IGY because scientists had predicted it would be a time of heightened sunspot activity. Nobody knows exactly what causes sunspots, but they frequently occur together with explosions of hydrogen gas on the surface of the sun. Particles of hydrogen from such eruptions are electrically charged, and when they collide with the ionosphere—the earth's outer atmosphere—they trigger magnetic storms, which disrupt radio communications and interfere with magnetic compasses. The magnificent aurora that ushered in the IGY was an early indication that the prediction of intense sunspot activity was correct. As hoped, the IGY scientists gleaned countless observations of the results of solar activity.

But the scientists could not predict just how long-lasting and far-reaching an enterprise they were launching in the IGY. When the 18 months were over, the Antarctic investigations had yielded much more than expected, including a picture of many of the continent's visible surface features, half of which had been unknown as the IGY began.

A welcoming committee of Adélie penguins totters past a phlegmatic seal toward the research ship *Polarsirkel*, temporarily halted in the puck ice off Antarctica in 1980. The vessel surveyed almost 1,000 miles of coast in search of a site for a new West German research station.

The scientists discovered that, isolated though it seemed, Antarctica affects and is affected by activities in the far reaches of the globe. They found that the blanket of glacial ice holds secrets to ages past, reacts to the undertakings of modern man and even reveals clues to the future of the earth.

The work was so fruitful and the cooperation so beneficial that the scientists agreed to continue their research beyond the period of the IGY. That decision was made internationally binding in 1961, when the 12 nations they represented ratified the Antarctic Treaty—the first instrument in the history of mankind to devote an entire continent to the pursuit of scientific knowledge.

To be sure, the scientific undertakings were tinged with political motives. Between 1908 and 1946, seven countries had made territorial claims on Antarctica, either on grounds that they had sponsored early expeditions

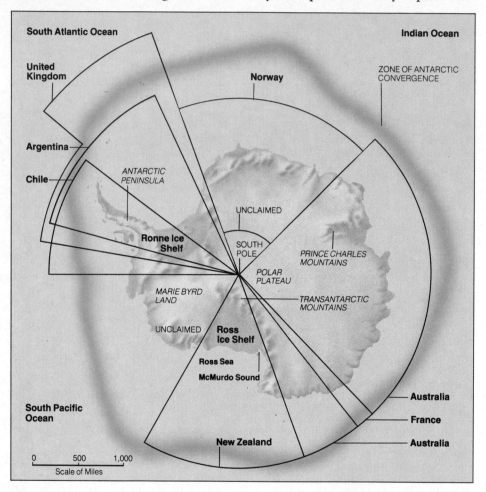

Antarctica resembles a much-divided pie in this map showing the territorial claims of seven nations. A 1961 treaty suspended all such colonial pretensions until 1991, but several countries continued to bolster their potential claims by maintaining research stations there.

to the continent, as had Great Britain, France, Norway, Australia and New Zealand, or on grounds of proximity, as asserted by Argentina and Chile. Other nations that had engaged in exploration—the United States and the Soviet Union, for instance—made no claims and refused to recognize those of others. The Treaty temporized on the territorial issue; it simply suspended all claims for a period of 30 years. But it specified that Antarctica was to be used for peaceful purposes only, and it stipulated that the findings of all scientific observations were to be made freely available throughout the world. Nine other nations acceded to the Treaty after 1961, and two of them, West Germany and Poland, joined in the research. By 1980 the number of research stations stood at about 40, the number of people conducting annual studies of the frozen continent had risen to 2,000 and the number of scientific disciplines under study had expanded to include such fields as biology and human medicine.

Perhaps the best example of the size and scope of the research activities in Antarctica as the 1980s began was the American base on Ross Island at McMurdo Sound, headquarters for American operations on the continent. It had a summer population of about 800 and a physical plant of 100 buildings. Most were plain and functional, but the administration building had been adorned in the style of a Swiss ski chalet. The phenomena studied there ranged from respiratory ailments (hardly anyone contracts ordinary illnesses in the nearly germ-free cold, but a medical team deliberately introduced respiratory viruses to volunteers among the base personnel) to the effects of variations in the earth's magnetic field, as detected by satellites in orbit about 600 miles up.

By and large, Antarctic research stations lie in sectors where particular nations have vested interests. But the scientists frequently mount cooperative undertakings and routinely share their findings. From their base about 1,000 miles inland from Victoria Land, the Soviets send regular reports to world meteorological centers on their studies of atmospheric physics. Teams from the United States, New Zealand and Japan have worked together in exploring and studying the ice-free valleys *(pages 158-159)*. New Zealand teams make topographical surveys and maps for the use of all students of the continent, and Australia often provides support for field parties, no matter what flag they fly.

151

At the South Pole the various national sectors converge at a ceremonial orange-and-black-striped pole surrounded by flags representing the nations that accede to the Antarctic Treaty. About 300 feet away from the pole and the flags stands an American base that is the continent's most celebrated installation. Sheltered against the wind by a 52-foot-high aluminum geodesic dome, three prefabricated buildings provide sleeping accommodations for 40 and amenities that would boggle the minds of Robert Scott and his contemporaries—hot showers, lavish rations that include Rock Cornish game hen with sherry, and ham radio equipment to keep the residents in touch with home. From their amenable South Pole base the Americans monitor the atmosphere for carbon dioxide and other pollutants, observe auroral displays, study solar energy and how it affects the

With an elaborate crevasse detector affixed to its front, a tractor creeps across a snowfield in Antarctica in 1960. Designed to sense a change in the electrical properties of the underlying snow, the device failed to distinguish quickly between a snow bridge over a yawning crevasse and a lightly packed but safe layer of snow.

polar ice, and record the effects of isolation on the people who spend the sunless winter there.

For all the advances in facilities and equipment, the fundamental operation of Antarctic research—the field trip—remains rigorous in the extreme. Scientists arriving on the continent are provided with a survival manual that gives instruction in such arcane matters as how to make an emergency shelter out of snow, how to use a mirror for signaling and how to cook with airplane fuel. Americans are urged to purchase high-risk life insurance policies and to make out their wills. Nevertheless, field trips are essential. The scientists must journey into the Transantarctic Mountains to explore the valleys and glaciers, and to the top of Mount Erebus to monitor its volcanic activity and study its lava; and they need to make frequent expeditions onto the Ross Ice Shelf and to the polar ice sheets to examine the ice itself.

During the era of intense scientific activity inaugurated by the IGY, territorial issues have from time to time reasserted themselves. But the spirit of cooperation has prevailed, and the glacial continent of Antarctica has continued to be a laboratory for all the world.

From the beginning, the omnipresent snow and ice have inevitably been the object of extensive investigation. Both are analyzed for their make-up (the elements that are part of their structure, and the chemicals and microparticles that come from the atmosphere) and for their dynamics (how the ice flows in various locations and how it interacts with the atmosphere).

One of the earliest methods for studying the snow was to dig pits and examine the walls for clues to the age of the layers. Another was to cut out three-cubic-foot blocks, then slice them into one-inch-thick sections that could be conveniently scrutinized for annual layering. A more sophisticated and far more complicated method is to drill into the ice, a practice that is undertaken sometimes to find out what lies at the bottom, sometimes to measure thickness and temperature, and sometimes to extract cores for later examination. Drilling remains a tricky business even after more than two decades of experience. In the upper layers the ice shatters easily. Several hundred feet down, it is so plastic and mobile that the hole may close up quickly, as an American drilling crew found out in 1976, when trying to

After the electric crevasse detectors proved fallible, tractor drivers had to trust their own eyes to avoid the slight surface depressions that signal the presence of a snow bridge. Mishaps continued to occur, however. Here, a tractor whose operator failed to spot the subtle cue founders in a crevasse in Enderby Land.

penetrate the Ross Ice Shelf. After 16 days, the drill had reached a depth of about 1,100 feet. With scarcely 300 feet to go, the work stopped for a regular change of crews. When the new shift came on half an hour later, the ice had crept forward and trapped the drill at the bottom of the hole. No tool the men had on hand could extract it or get it going again, and they had to abandon the project until the following season.

When the drillers returned to the site in December of 1977, they brought with them a thermal drill, developed for quarrying granite, which generated a supersonic jet of hot gas. The new device burned through the 1,375-foot ice shelf in nine hours. For the first time, scientists were able to examine the sea beneath the shelf; a team of biologists lowered a television camera and caught a glimpse of primitive crustaceans thriving in the darkness and the cold. The sea floor was dredged for samples and yielded pollen grains estimated to be 14 to 20 million years old. The ice kept creeping all the while, however, and every three days the hole had to be reopened.

Most frequently, the aim of Antarctic drilling is to obtain a series of cylindrical cores that together provide a continuous vertical record of the ice sheet. The procedure involves drilling about 20 feet, extracting a core, drilling another 20 feet, extracting the second core, and so on. The work has to be done with infinite care, because the information the glaciologists are seeking can be obliterated if the core is crushed, broken or even scratched. Continuous records of remarkable length have been assembled.

In 1968 at the American station in Marie Byrd Land, scientists obtained a complete vertical record of ice more than a mile thick; the ice at the bottom of the sheet there is estimated to be 50,000 years old. Researchers aspire to assembling a continuous core from the heart of East Antarctica, where the ice is nearly three miles deep and could provide a record dating back perhaps half a million years. But the state of drilling is not yet up to such an undertaking.

Still, any number of individual cores have been retrieved from sites at the South Pole, on the Ross Ice Shelf, in Marie Byrd Land, in the Ellsworth Mountains just south of the Antarctic Peninsula and on the ice sheet east of the Victoria Land mountains. The cores are packed, section by section, in dry ice and transported to laboratories in the participating nations. The State University of New York at Buffalo has an enormous collection—more than 23,000 feet of cores from several sites in both Antarctica and Greenland.

After more than a decade, scientists have yet to exhaust the wealth of information offered by the particles entombed in the ice cores—ash from prehistoric volcanic eruptions, cosmic dust wafted to the earth from space, nitrates that shower down from the atmospheric disturbances accompanying auroral activity and thus leave a record of sunspot cycles, salts from sea spray stirred up by storms and borne inland by wind, air bubbles that show the

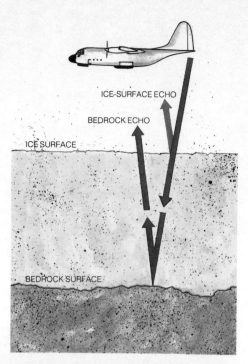

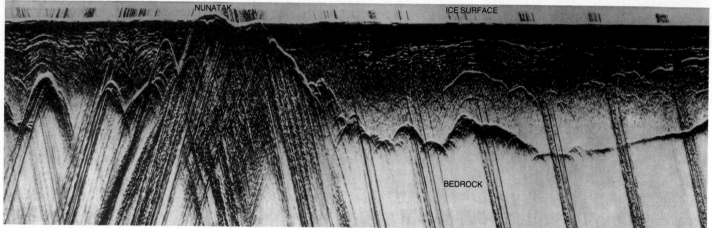

composition of the ancient atmosphere and the degree of modern pollution. Remote though it is from the industrial world, Antarctic ice has been found to contain traces of lead from automobile exhaust fumes, of the insecticide DDT and of radioactive fallout from the peak period of atomic bomb testing during the 1950s.

A major concern of the scientists who examine ice cores is climate, and the ice is full of clues. One of the most informative is an infinitesimal variation in the molecular structure of water, which contains two different isotopes, or atoms, of oxygen: oxygen 16, so-called because the nuclei contain eight neutrons and eight protons; and oxygen 18, containing two extra neutrons. Other factors being equal, snow that forms when it is relatively warm contains more of the heavier oxygen atoms than does snow that falls during a cold period. The ratio of O-18 to O-16 atoms can be measured in a laboratory by a mass spectrometer, which measures the weight of the atoms of oxygen. From the results, scientists can tell the difference between summer and winter snow, and they can discern long periods during which the global climate was generally warm or generally cool. It is known from other evidence that during the past four billion years, there have been at least four periods of millions of years' duration during which ice has covered major portions of the globe and global temperatures have been well below normal.

A low-flying aircraft (*diagram, top*) transmits radio signals down through the Antarctic ice and records the echo from the underlying rock. The result is a continuous image of the icebound continent's profile—in this case, a range of buried mountains barely cresting the glacier.

And the pattern of temperature changes during the most recent ice age has been found in the ice cores.

In addition to pondering the structure of ice, scientists want to know how thick the Antarctic sheet is and what lies underneath it. At the time the IGY began, the method generally used was seismic: The depth of the ice was determined by setting off explosions of TNT or dynamite on the surface, and timing the return of the shock waves from the underlying bedrock. From their various posts, the Russians, the Americans and the British ranged far afield, their tractors hauling trains of sledges loaded with tens of tons of fuel, food and provisions, setting off explosive charges every few miles. The work was slow and hazardous; many a tractor pitched into a crevasse, and because of the delays and the need for caution, only a few such soundings could be made per day. Furthermore, since the soundings were done only every few miles, the record obtained was not continuous.

A radical advance in charting subglacial topography came after it was discovered in the 1950s that ice was transparent to radar. The discovery was not a happy one; it came after several aircraft had flown into the Greenland ice sheet and crashed because the on-board radar altimeters were telling the crews how far they were above the bedrock, not the ice surface. By 1969, airborne radars recording their soundings on film or tape were providing continuous profiles of the Antarctic continent. And by 1980, seismic and radar soundings together had mapped more than 4.5 million square miles of ice-covered terrain.

The hidden landscape revealed by these and other techniques is a remarkable one of mountains, troughs, tablelands, great sedimentary basins, and even lakes of meltwater, some of them more than 100 miles long. Two distinct geological regions have been found, one on either side of the Transantarctic Mountains. The mountain chain itself as well as the land to the east of it consists of a platform of rock formations that are 600 million to 2.8 billion years old. West of the Transantarctic Mountains most of the continent actually lies below sea level (though the ice prevents the incursion of sea water) and is from 50 to 200 million years old.

The emerging picture of the land beneath the ice also yielded unexpected confirmation of the theory of continental drift. The new data revealed that East Antarctica strongly resembles, in structural and geological detail, Australia, South Africa and peninsular India, whereas West Antarctica is similar to South America. In fact, the narrow mountain chain that forms the backbone of the Antarctic Peninsula appears to have once been connected to the Andes.

These similarities added even more weight to the notion that continents move, first articulated in 1912 by the German scientist Alfred Wegener— a meteorologist and astronomer who dared to tread on the geologist's domain. Looking with a fresh and unspecialized eye at similarities in the opposing coasts of South America and South Africa, and in widely distributed fossils, Wegener discerned that many landforms appear to be direct continuations of others. In a remarkable leap of imagination, he proposed that the continents long ago were rent asunder from one primordial mother continent and that—like icebergs calved from a glacier—they drifted apart until they arrived at their present positions. He named his hypothetical mother continent Pangaea, Greek for "all earth." The thought was so radical that, when Wegener died in 1930, scientists who accepted his theory were still a distinct minority.

The idea suddenly took on new life in the early 1960s. By that time oceanographers were discovering on the floors of the Pacific and Atlantic

Oceans evidence that the sea floor is slowly spreading. From this finding and others grew the revolutionary concept of plate tectonics, the theory that the earth's crust consists of half a dozen or more separate, constantly moving plates that originate as molten matter welling up from below at the midocean ridges.

Just as the earlier topographical studies in Antarctica had supported the idea of continental drift, so the further findings of recent years have advanced the theory of plate tectonics. In 1960, a rocky outcropping of the Transantarctic Mountains southeast of the Beardmore Glacier was found to contain a 900-foot-thick deposit of tillite—glacial till compacted into rock. The tillite, 250 million years old, and the overlying sandstone, laced with beds of coal and fossilized fern, resembled, layer for layer, tillite and sandstone found in South Africa, India, Australia and South America. In December 1967, a sandstone bluff atop some 230-million-year-old sedimentary rocks in the Transantarctic Mountains yielded fragments of bone thought to belong to the lower jaw of a labyrinthodont, a salamander-like amphibian whose remains are found in the sandstone of South America, Australia and South Africa. Two years later a team of geologists found in a stretch of exposed sandstone 400 miles from the South Pole a tusk and part of a jawbone of a 200-million-year-old creature named lystrosaurus, a stubby, four-legged reptile known to have flourished in South Africa and peninsular India. And in March 1982, still another group of scientists announced the first discovery in Antarctica of a land-mammal fossil—a three-inch jawbone and a couple of teeth of a 40-million-year-old rat-sized marsupial previously known to have ranged in South America and Australia.

None of these creatures could possibly have swum to Antarctica across thousands of miles of open water. Together with the coal, rock formations and fossil plant life found throughout Antarctica and the Southern Hemi-

Beneath the smooth dome of the miles-thick Antarctic ice sheet lies the rugged and varied terrain revealed by radio echo soundings and illustrated here. Some of the Transantarctic Mountains soar higher above sea level than the Rocky Mountains in North America, while the lowlands of West Antarctica are as much as 8,200 feet below sea level.

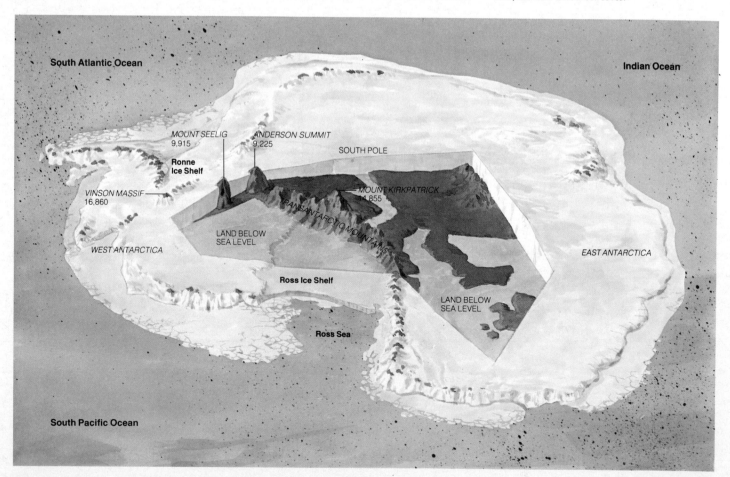

sphere, their bones have convinced virtually all scientists of the validity of the theory of plate tectonics.

The breakup of the mother continent is now thought to have taken place over a period of many millions of years, beginning about 180 million years ago. After South America parted from South Africa, India and Australia migrated northward, leaving Antarctica. There, about 16 million years ago, glaciation began, eventually covering the continent.

The discovery of Antarctica's similarities to South Africa and South America immediately raised the possibility that beneath and around the forbidding ice there was an El Dorado of minerals and fossil fuels. Recent probes support the possibility. A 1,500-mile strip of terrain along the east side of the Transantarctic Mountains embraces a coal field that may prove to be the largest on earth. In the Prince Charles Mountains near the coast facing the Indian Ocean lies a 75-mile-long, 330-foot-thick deposit of iron ore, enough to meet present world demand for 200 years. In the Dufek Massif southeast of the Antarctic Peninsula, airborne detection systems have mapped a 20,000-square-mile area of rock that closely resembles a region in the South African Bushveld where deposits of platinum, chromite, tin, magnetite, nickel, vanadium and fluorite yielded estimated revenues of $1.3 billion in 1980. And in ice-free outcroppings elsewhere geologists have found small but tantalizing amounts of several other minerals, among them zinc, lead, copper, nickel, cobalt and uranium.

The most lucrative prize of all may be oil. In the continental shelf under the Ross Sea, geophysical surveys have detected thick sedimentary rock formations of a sort frequently associated with petroleum deposits. Traces of ethane and methane discovered during exploratory drilling also suggest the presence of oil. How much petroleum is anyone's guess, but perhaps tens of billions of barrels of it.

Inevitably, the presence of such resources has kindled flames of commercial interest. Even before the end of the 1960s, business organizations had asked the governments of Britain, the United States, Australia and New Zealand for rights to search for oil—and the requests raised questions of proprietorship. In general, the nations that claimed territorial rights also claimed ownership of any wealth lying beneath the ice, while the governments that made no claims argued that the riches, like scientific information, should be available to all. Since the Antarctic Treaty of 1961 suspended all claims for 30 years, the question is moot for the time being.

But most nations—including those that disclaim territorial ownership—continue to engage in subtle but insistent assertion of their presence on the continent. Argentina, Australia, Chile, France, New Zealand, Norway and the United Kingdom all issue postage stamps proclaiming jurisdiction over their sectors of the continent. Chile has ostentatiously staged presidential visits to the Antarctic Peninsula, held Cabinet sessions there and passed a law requiring Chilean publishers of atlases to include the national claim to Antarctica on all maps. Argentina has been even more imaginative in asserting its claim; it encourages weddings at the Argentine research stations on the Antarctic Peninsula and in 1977 went so far as to transport a pregnant woman to one such station, where, a few months later, she gave birth to the first baby ever born in Antarctica.

Other nations, while making no territorial claims, have managed to communicate a keen, and not always exclusively scientific, interest in the continent. In the mid-1970s the two latecomers to the Treaty fraternity joined in the scientific research: Poland established a research station off the Antarctic Peninsula for the investigation of marine life, and West Germany began oceanographic studies in the Weddell Sea off the Ronne Ice Shelf. By 1980

The Sere Oases of an Icy World

For many years, explorers assumed that the surface of Antarctica was one enormous sheet of ice, broken only here and there by rock outcroppings or mountain peaks. Thus, when British explorer Robert Falcon Scott and his men in 1903 came upon a 25-mile-long valley that contained almost no glacier ice, their astonishment knew no bounds. "It seemed almost impossible," Scott wrote, "that we could be within a hundred miles of the terrible conditions we had experienced on the summit."

Later explorers found two similar valleys nearby, and the surrounding land, for a space of about 2,000 square miles, also had an unusual shortage of ice *(map)*. The explanation proved to be straightforward. More than four million years ago, glaciers flowing through the valleys were cut off from the ice sheet that fed them by the slowly rising Transantarctic Mountains. Gradually the isolated glaciers evaporated, leaving the valleys almost ice free. Since then, cold, dry winds from the Polar Plateau have continually swept away the meager three inches of snow that falls in the valleys each year, and have eroded their rocks into fantastic shapes.

Only a few glacier snouts now reach into the valleys, feeding small lakes with their scant meltwater. In the lakes, scientists have found something even more surprising than the valleys themselves. In 1981, under the ice covering Lake Fryxell, biologists discovered an extensive mat of living blue-green algae *(right)* almost identical to the algal mats that dominated the world's waters three billion years ago. By studying these living replicas of an ancient ecosystem, biologists hope to gain new insights into the early history of life on earth and the adaptations of simple organisms to frigid temperatures and prolonged periods without sunlight.

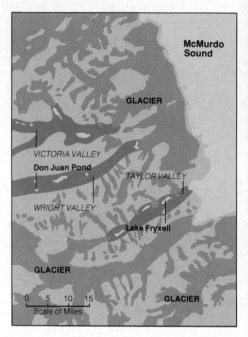

The ice-free valleys of Antarctica's Victoria Land *(brown)* and the exposed region around them *(tan)* occupy a tiny but scientifically significant corner of the ice-covered 5.5-million-square-mile continent.

A mat of blue-green algae, illuminated here by artificial light, blankets the bottom of Lake Fryxell in the Taylor Valley.

The snout of Victoria Upper Glacier extends only slightly into Victoria Valley, a U-shaped trough sculpted by a larger glacier more than four million years ago. The dark stratum on the mountain in the background was formed as molten lava solidified about 180 million years ago, when the supercontinent of Gondwanaland began to break apart.

An ornately carved boulder at the edge of Don Juan Pond bears witness to the abrasive power of the harsh winds that sweep into Wright Valley from the Polar Plateau. Particles of chloride and other salts, dissolved and eroded from the nearby rocks, have made the pond so salty that it never freezes.

the Soviet Union had expanded the number of its year-round stations in various sectors of the continent from two to seven. Representatives of several American oil companies have served on the U.S. State Department's Antarctic Advisory Committee. And Japan's National Oil Corporation is using a highly sophisticated seismic technique to explore the bottom of the Weddell Sea. The corporation does not pretend to be engaged in scientific research: It is prospecting for oil, and plans to be in position to seize the spoils should the time ever come.

Such jockeying for position may be premature, for the practical problems of securing the continent's riches are immense and far from solved. Most of the potentially mineral-rich Dufek Massif lies beneath nearly a mile of ice, and even if the technology existed for extracting ores through such a moving barrier, the labor of mining in the savage Antarctic climate would sorely try human endurance. The prospects for tapping the hypothetical offshore oil deposits are equally remote. By the 1980s, plans existed for such technological marvels as drills that could penetrate to a depth of 16,500 feet and submarine storage tanks that could be moored deep enough to be beyond the reach of icebergs. But none of these schemes had yet been attempted even on a trial basis. And no tanker could guarantee safe passage of the cargo; to get the oil to markets thousands of miles distant would require the ships to pass through the storm-tossed latitudes between 60° and 40° S. Edward P. Todd, Director of the U.S. National Science Foundation's Division of Polar Programs, sums up the commercial prospects by saying, "It will be a long time before anyone makes a dime on Antarctic minerals."

The Antarctic studies in oceanography and biology have, however, turned up one exploitable resource in the teeming marine life that thrives in the frigid waters surrounding the continent. In seeming paradox, the Ant-

Two American scientists examine layers of snow in the wall of a pit dug in Antarctica in 1958. Analysis of core samples helped them to reconstruct annual snow accumulation on the ice sheet over several decades.

arctic ice that makes life on the continent virtually impossible does not have the same effect on the nearby sea. The reasons are complex and have to do with the mixing that takes place at the Antarctic convergence—the zone between 50° and 60° S., where the warm, saline currents of the Atlantic, Pacific and Indian Oceans meet the cold, oxygen-rich waters flowing north from the shores of Antarctica. The collision of all those currents causes a monumental turbulence that stirs up the phosphates and nitrates emanating from the decomposing bodies of fish, plants and animals on the ocean floor. When they reach the surface, these nutrients foster the growth of phytoplankton, microscopic one-celled floating plants. The plants, in turn, provide plentiful food for marine creatures.

The most abundant form of sea life south of the Antarctic convergence is a two-inch reddish, shrimplike crustacean known to scientists as *Euphausia superba* and to fishermen as krill, a Norwegian whaling term that translates loosely as "small fry." Krill are a vital link in the marine food chain; they eat phytoplankton and provide sustenance for no fewer than 31 species of whale, seal, squid, bird and fish.

Krill are easy to net. They are visible because they congregate on or near the surface in schools so dense that they make the water seem bright pink by daylight and phosphorescent blue-green by night. The distribution varies, but some schools yield 35 pounds of krill per cubic yard.

Trawlers of half a dozen nations harvest krill from Antarctic waters for both human and animal consumption. In Japan the krill are marketed whole as seafood and are generally regarded as a suitable substitute for the native species of shrimp. In the Soviet Union, processed krill are used to fortify sausage and cheese spread; in Poland, West Germany and Chile they are turned into feed for cattle and poultry.

Between 1978 and 1981 the harvest of krill leaped from 122,000 tons to

Wearing a sterilized uniform, special gloves and a surgical mask, a British chemist saws through blocks of snow quarried from the Antarctic ice sheet. At university laboratories in England and Belgium, the samples were screened for traces of atmospheric pollutants, including strontium 90, lead and carbon monoxide.

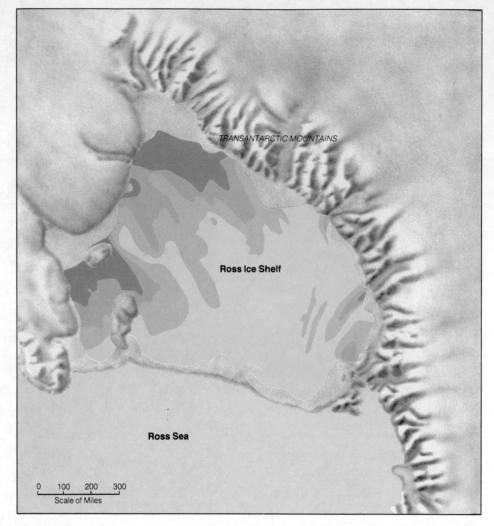

975-1,300 FEET

1,300-1,625 FEET

1,625-1,950 FEET

1,950 FEET AND OVER

TRANSANTARCTIC MOUNTAINS

Ross Ice Shelf

Ross Sea

0   100   200   300
Scale of Miles

A map of Antarctica's Ross Ice Shelf shows the striking variations in the thickness of the ice found wherever radar readings have been made. Because each of the 18 major streams of ice that feed it flows at a different rate, the thickness of the 216,000-square-mile ice shelf ranges from about 975 to 2,600 feet.

about 200,000 tons, and the figure is likely to go higher as the attributes of krill become more widely known. Krill are probably the world's largest untapped source of protein. They contain as much protein by weight—15 per cent—as beefsteak. Some scientists say annual harvests of up to 150 million tons can be taken without danger to the marine food chain because krill reproduce themselves at a rate of 150 million tons every year; others worry that heedless exploitation will deliver krill to the fate of the whales and the seals, which were hunted almost to extinction in the 19th Century. In 1980 the Antarctic Treaty nations considered regulating the catch, but could not agree on what limits to set.

The lure of worldly riches in the Antarctic continent and its surroundings has not distracted scientists from the ice itself, with its store of information about the history and evolution of the earth and its climate. "Climatic research is bringing the polar regions into a new perspective," says William W. Kellogg, a climatologist at the National Center for Atmospheric Research. Both polar regions, he points out, are sensitive indicators of climatic change because their large masses of ice "respond to temperature changes in a special way." The ice also causes temperature changes; it reflects back into space most of the radiation it receives from the sun before atmospheric warming can take place, and it chills the warm winds and ocean currents that flow south from the temperate regions. The greater the area of the ice cover, the more heat will be dissipated into space. The polar ice therefore plays a significant role in the giant thermodynamic machine that governs climate all over the globe.

Many scientists have sought to find in the Antarctic ice sheet an explanation for the ice ages of the past—and through such an explanation, to anticipate the future. Among the most provocative theories is one proposed in 1964 by New Zealand glaciologist Alex T. Wilson, who described the waxing and waning of ice ages as a cyclical occurrence. In his view, the Antarctic ice sheet periodically becomes uncoupled from its bed and surges ahead—in a grand variation on the performance of surging valley glaciers, which often flow at many times their normal rates in Alaska *(page 62)* and elsewhere. Wilson postulates that if the Antarctic ice sheet surged, the resulting frictional heat would accelerate its movement by increasing the amount of lubricating meltwater at its base. The surging ice sheet, he thinks, could expand over the sea until it eventually covered an area of perhaps 15 million square miles, or three times the area covered by the present ice sheet.

That much ice, reflecting 80 per cent or more of the sun's radiation, would decrease the amount of solar radiation available to warm the earth's atmosphere by as much as 4 per cent. In Wilson's scenario, the resulting drop in average temperature, though only a few degrees Fahrenheit, would

The dynamics of Antarctica's massive ice shroud produce singular surface features: An ice rise *(right)* protrudes from the surface where a section of the ice shelf has run aground on a shoal; an ice rumple *(below)* forms where the ice shelf is sliding over an obstacle and the moving ice is fractured and compressed.

A field biology camp on the Ross Ice Shelf near Antarctica's McMurdo Sound is surrounded by swirls of vehicle tracks on the otherwise pristine landscape. In 1981, scientists at this outpost studied the behavior of Weddell seals.

be sufficient for ice sheets to form all over the Northern Hemisphere and grow to a total area of eight million square miles. The glaciation would further decrease the warming effects of solar radiation and would eventually cause a worldwide temperature drop of nearly 11° F.

The dramatic expansion would also cause the runaway ice sheet to thin out, and at some point the frictional heat generated would dissipate and ice would begin to accumulate again, and freeze to the bedrock. The surge would end, and the ice sheet would return to creeping at its former pace of only inches per day.

Meanwhile, because the ice sheet could no longer be replacing ice calved off in icebergs, the ice shelf extending over the ocean would diminish in size. More and more of the water of the southern ocean would be exposed to sunlight, become warmer and gradually distribute the warmth to the atmosphere of the entire earth. The ice sheets of North America and Europe would melt away—and the ice age would come to an end.

Many scientists have added theoretical detail to Wilson's idea of a surging ice sheet. Glaciologists George Denton and Terence Hughes of the University of Maine believe that a cycle such as the one Wilson describes may now have passed the stage of maximum expansion. They have gathered data suggesting that the West Antarctic ice sheet may once have been three times bigger than it is at present and that it is now in an advanced state of weakness.

No fewer than 19 ice streams drain seaward from the ice sheet at a rate of five to 10 feet a day. Denton and Hughes believe that these currents of ice would flow even more rapidly, with dramatic consequences for the ice sheet, if they were not restrained by two major ice shelves, the Ross and the Ronne. The shelves, in the view of Denton and Hughes, serve as plugs that stop the ice sheet from draining away altogether.

What might happen without such restraint is indicated by the Pine Island Glacier, an ice stream that flows unimpeded into the Amundsen Sea, about midway between the Ross and Ronne shelves. The Pine Island is one of the world's fastest glaciers; it travels 33 feet per day—more than three times the speed of the streams that are dammed by the ice shelves. Denton and Hughes believe that the Pine Island Glacier did not always behave as it does now. Glacial deposits on the sea floor for some distance beyond its

Engineers lower a hot-water drill into the Ross Ice Shelf in 1978. With water that had been heated to more than 200° F., the drill cut a hole three feet in diameter through more than 1,300 feet of ice in about 11 hours.

present terminus indicate that perhaps as recently as a century or two ago an ice shelf may have existed there. The disappearance of the postulated ice shelf could have pulled the plug on the Pine Island Glacier and allowed it to drain ice at its present rapid rate.

Denton and Hughes express concern that the Ross and the Ronne Ice Shelves may also be in danger of disappearing. Except in a few places where these shelves are anchored on islands or underwater shoals, both of them are floating in the sea. In places where the ice meets no resistance from bedrock, it flows more rapidly—in some areas at the rate of as much as eight feet a day—and the ice stretches and thins out. At their terminal edges, the Ross and the Ronne Ice Shelves are only a few hundred feet thick—a fraction of their thickness several hundred miles closer to the shore, where they are grounded on bedrock.

These shelves calve off most of the great tabular icebergs of the southern seas. Denton and Hughes suggest two things that might speed up the calving process and eventually unleash the fast-moving ice streams. One of these would be an increase in average temperature, which would cause the surfaces of the shelves to melt; the meltwater would pour down into stress cracks and weaken the shelves until they eventually split asunder. The other would be a rise in sea level, perhaps from the melting of glaciers in the Northern Hemisphere, which would lift the ice shelves off the anchoring bedrock and cause the entire slab of ice to stretch out more rapidly. In either case, the rate of calving would eventually exceed the supply of new ice reaching the shelves. The shelves would then shrink until they completely disappeared. Without them the ice streams would flow directly into the sea and contribute to a further rise in global sea level.

In the normal course of nature's long-term cycles of warming and cooling, there would seem to be no imminent danger of widespread deglaciation. But nature is no longer being allowed to take its course; modern civilization affects the global climate on a scale hitherto unknown in history. The century that began in 1876 is believed by some scientists to have been the warmest in approximately 4,000 years, and most of them attribute that warmth to

The variety of novel experiments made possible by the piercing of the Ross Ice Shelf is illustrated here. Geologic specimens were taken from the seabed, bottles and nets captured water samples and exotic marine animals, and a TV camera peered into the frigid gloom.

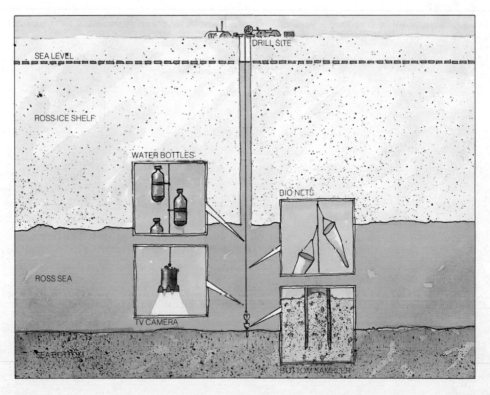

165

The flags of the nations that accede to the 1961 Antarctic Treaty stand in ceremonial array in front of the Amundsen-Scott Station at the South Pole. The 52-foot-high aluminum geodesic dome at this American outpost can accommodate as many as 40 residents.

the vast amounts of carbon dioxide that are being pumped into the atmosphere by the industrial nations in their burning of the fossil fuels coal, oil and gas. Carbon dioxide influences climate because, as was first theorized by British scientist John Tyndall in 1861, it permits incoming solar radiation to pass freely through but blocks a good deal of the heat being radiated back from the earth's surface. That phenomenon has been dubbed the greenhouse effect because the glass in a greenhouse admits sunlight and retains warmth.

In the century following Tyndall's observation, the rate of consumption of coal, gas and oil increased enormously with spreading industrialization. By the start of the 1980s, the amount of carbon dioxide being released into the atmosphere every year—approximately 20 billion tons—was increasing at an annual rate of 4.3 per cent. Scientists predict that if consumption of fossil fuels continues to rise as it has been doing, the next 50 years will bring a doubling of the levels of atmospheric carbon dioxide—and consequently a rise in average global temperatures of 4° or 5° F., and a rise in the polar regions of 8° to 11° F.

For the ice sheet lying over West Antarctica, such a warming could have radical consequences. "Once this comparative level of warmth had been reached," says John H. Mercer of the Institute of Polar Studies at Ohio State University, "deglaciation would probably be rapid—perhaps catastrophically so." Mercer believes that with the exception of the ice that is now grounded above sea level in the mountains, the entire West Antarctic ice sheet could vanish in less than a century's time. To support this assertion, he cites the example of the North American ice sheet, which some 8,000 years ago was centered over what is today Hudson Bay. With the onset of a global rise in temperatures, that ice sheet disappeared completely in less than two centuries.

Scientists differ on which, if any, of the foregoing theories is correct. But one matter on which they have no quarrel is that if the West Antarctic ice sheet were to vanish as a result of any process, the effects would be noticeable all over the world and might be disastrous in some places. Sea level would rise at least 20 feet, submerging land in low-lying areas such as Florida, Louisiana and Delaware. New York, Tokyo, London, Amsterdam and other great coastal cities would face the danger of severe flooding.

No one knows for certain, of course, that melting of the ice shelves would actually result in a collapse of the ice sheet in West Antarctica; some other process not yet understood might come into play to forestall that eventuality. Even those who foresee collapse believe that it might not happen as fast as Mercer predicts, but might take place over a period of several centuries. "About 16 'mights' in a row have to be piled up," says University of Wisconsin geophysicist Charles Bentley. "There are fascinating possibilities, some of them dramatic, but there is no proof that anything dramatic will happen."

Nor does anyone know that the climate is irreversibly warming. "Put any two climatologists together," says Murray Mitchell of the National Oceanic and Atmospheric Administration, "and that's a subject certain to start a fight." At the very least, however, Mercer's theory of the rapid disappearance of the West Antarctic sheet points to the need for better scientific understanding of the greenhouse effect and of the complex interaction between ice and climate. The entire industrialized world depends at present on burning fossil fuels, and the resulting carbon dioxide dissipates so slowly that it could endure for thousands of years, bringing a millennium or more of warmth such as the earth has not experienced in the last million years.

The mere consideration of such a prospect suggests that Antarctica's most valuable resource may prove to be neither oil wells, nor veins of minerals, nor harvests of protein-rich krill, nor any other material commodity. For scientists, at least, the greatest treasure to be found on the ancient continent, with its majestic ice inching slowly into the sea to take its place in the world's hydrocycle, is a wealth of knowledge about the history and dynamics of the entire planet. **Ω**

The Transantarctic Mountains, rooted in the
vast Polar Plateau at the center of the continent,
lift their peaks above a sea of ice.

PICTURE CREDITS

The sources for the illustrations that appear in this book are listed below. Credits for the illustrations from left to right are separated by semicolons, from top to bottom by dashes.

Cover: Bob and Ira Spring. 6, 7: Hiroshi Hamaya, Fujisawa, Japan. 8, 9: Chris Bonington from Bruce Coleman, Ltd., Middlesex, England. 10, 11: © Tom Bean 1981. 12, 13: © Gary Martin from Bruce Coleman, Inc. 14, 15: © Georg Fischer/VISUM, Hamburg. 16: Geodetic Institute, Copenhagen. 18: Art by Richard Schlecht. 20: John E. Fletcher © National Geographic Society. 22: J.-M. Biner, courtesy Musée de la Majorie, Sion, Switzerland. 23, 25: Library of Congress. 26: From *Studies on Glaciers preceded by the Discourse of Neuchâtel* by Louis Agassiz, Hafner Publishing Company, New York and London, © 1967 Office of the Rector, University of Neuchâtel, Neuchâtel, Switzerland. 28: From *Louis Agassiz: His Life and Correspondence,* Vol. 1, by Elizabeth Cary Agassiz, the Riverside Press, Cambridge, Massachusetts, 1885. 29: From *The Life and Correspondence of William Buckland* by Mrs. Gordon, London, 1894. 30: Map by Bill Hezlep. 32: Art by Richard Schlecht. 34: Art by I'Ann Blanchette. 35: John Cleare/Mountain Camera, London. 36: Norwegian Polar Research Institute, Oslo. 38-45: Vittorio Sella, Istituto di Fotografia Alpina, Biella, Italy. 46, 47: William O. Field, American Geographical Society. 48: National Park Service, Yosemite National Park Collections. 50: Art by Richard Schlecht. 51: Susan L. Herron. 52: Photo by Austin Post/U.S. Geological Survey (FR602560). 53: Art by Richard Schlecht. 54: Markus Aellen, Zurich. 55: Bjørn Wold, Oslo—art by Richard Schlecht. 56, 57: The Bondhus Project, Norway. 59: Art by Richard Schlecht. 60: E. W. Smith, Lancashire, England—Dr. James W. Fullmer—© Leo Touchet. 61: B. Ned Kelly, Bristol, England; E. W. Smith, Lancashire, England—David Falconer; E. W. Smith, Lancashire, England—Keith Gunnar from Bruce Coleman, Inc. (2). 62, 63: Austin Post/U.S. Geological Survey (middle, 70 N2-154 7.3.70). 64: Gunnar Hannesson, Reykjavik. 65: Art by Richard Schlecht. 66: Gunnar Hannesson, Reykjavik. 68-73: Art by Lloyd K. Townsend. 74, 75: Nicholas Devore from Bruce Coleman, Ltd., Middlesex, England. 76: © Geoff Doré from Bruce Coleman, Ltd., Middlesex, England. 77: © 1980 Jim Brandenburg from Woodfin Camp & Associates. 78, 79: Jeff Foott from Bruce Coleman, Inc.; Jerome Wyckoff. 80, 81: © 1981 R. Hamilton Smith; Jerome Wyckoff—EARTH SCENES/Breck P. Kent. 82, 83: Photo S.N.T.O., Zurich. 84: Derek Fordham from Susan Griggs Agency Ltd., London. 86: Steve McCutcheon. 87: Art by I'Ann Blanchette. 88: O. B. Olesen, The Geological Survey of Greenland, Copenhagen, Denmark. 91: U.S. Navy, courtesy Charles Swithinbank, Cambridge, England. 92: Art by Bill Hezlep. 93: © Bill Brooks from Bruce Coleman, Inc. 94: © The Illustrated London News Picture Library, London. 96: © Bruno Zehnder, New York. 97: James R. Holland © National Geographic Society. 98: UPI. 99: Harald Sund for *Life.* 100: © The Illustrated London News Picture Library, London. 102: Art by Richard Schlecht. 104, 105: Collection of Charles Swithinbank, Cambridge, England. 106, 107: © 1977 George Hulton from Photo
Researchers, Inc. 108, 109: Peter Johnson, Johannesburg. 110, 111: Francisco Erize from Bruce Coleman, Ltd., Middlesex, England. 112, 113: Eliot Porter. 114: Popperfoto, London. 116: Photo Bibliothèque Nationale, Paris. 118: The National Maritime Museum, London. 119: Mary Evans Picture Library, London. 121: Bildarchiv Preussischer Kulturbesitz, Berlin (West) — Frederic F. Bigio from B-C Graphics. 122: Popperfoto, London. 123: Royal Geographical Society, London. 124: Art by John Britt. 126, 127: Scott Polar Research Institute, Cambridge, England. 128: BBC Hulton Picture Library, London. 129: Popperfoto, London — BBC Hulton Picture Library, London. 130: Norwegian Polar Research Institute, Oslo. 132-135: Popperfoto, London. 136: RG 401/59 National Archives Gift Collection of Materials Relating to the Polar Regions, Laurence M. Gould Papers, BAEI-6. 138-147: Popperfoto, London. 148: Tobias Heldt for *Stern,* Hamburg. 150: Map by Bill Hezlep. 151, 152: U.S. Navy. 153: © Bruno Zehnder, New York. 154: Art by I'Ann Blanchette — Scott Polar Research Institute, Cambridge, England, and National Science Foundation. 156: Art by Richard Schlecht. 158, 159: Map by Bill Hezlep; Dr. John G. McPherson — Gordon Love; Kevin Schafer. 160: Emil Schulthess from Black Star. 161: Mark Vallance, Foolow, Derbyshire, England. 162: Map by Bill Hezlep and John Britt. 163: Collection of Charles Swithinbank, Cambridge, England. 164: Kevin Schafer. 165: National Science Foundation — art by Richard Schlecht. 166: National Science Foundation. 168, 169: Eliot Porter.

BIBLIOGRAPHY

Books

Adams, George F., and Jerome Wyckoff, *Landforms.* Golden Press, 1971.

Agassiz, Elizabeth, *Louis Agassiz: His Life and Correspondence,* Vols. 1 and 2. Houghton Mifflin, 1886.

Agassiz, Louis, *Studies on Glaciers.* Hafner Publishing Co., 1967.

Allison, Ira S., et al., *Geology: The Science of a Changing Earth.* McGraw-Hill, 1974.

Amundsen, Roald, *The South Pole: An Account of the Norwegian Antarctic Expedition in the "Fram," 1910-1912,* Vols. 1 and 2. London: John Murray, 1913.

Anderson, Madelyn Klein, *Iceberg Alley.* Julian Mesner, 1976.

Andrews, J. T., *Glacial Systems.* Duxbury Press, 1975.

Andrist, Ralph K., *Heroes of Polar Exploration.* American Heritage Publishing Co., 1962.

Armstrong, Terence, et al., *Illustrated Glossary of Snow and Ice.* Cambridge: The Scott Polar Research Institute, 1973.

Asimov, Isaac, *The Ends of the Earth.* Weybright and Talley, 1975.

Baird, Patrick D., *The Polar World.* Longmans, Green and Co. Ltd., 1964.

Barraclough, Geoffrey, ed., *The Times Atlas of World History.* London: Times Books Ltd., 1981.

Bates, D. R., ed., *The Earth and Its Atmosphere.* Basic Books, 1957.

Beaumont, Greg, *The Life of Glacier National Park.* National Park Service, 1978.

Billing, Graham, *South: Man and Nature in Antarctica.* University of Washington Press, 1964.

Birkeland, Peter W., *Putnam's Geology.* Oxford University Press, 1978.

Bixby, William, *McMurdo, Antarctica.* David McKay Co., 1962.

Bond, Creina, and Roy Siegfried, *Antarctica: No Single Country, No Single Sea.* Mayflower Books, 1979.

Brent, Peter, *Captain Scott and the Antarctic Tragedy.* Saturday Review Press, 1974.

Briggs, Peter, *Laboratory at the Bottom of the World.* David McKay Co., 1970.

Brunsden, Denys, and John C. Doornkamp, *The Unquiet Landscape.* Indiana University Press, 1978.

Byrd, Richard Evelyn:
Discovery: The Story of the Second Byrd Antarctic Expedition. G. P. Putnam's Sons, 1935.
Little America. G. P. Putnam's Sons, 1930.

Calder, Nigel, *The Weather Machine.* Viking Press, 1974.

Cameron, Ian, *Antarctica: The Last Continent.* Little, Brown, 1974.

Carter, Paul A., *Little America: Town at the End of the World.* Columbia University Press, 1979.

Chapman, Walker:
Antarctic Conquest: The Great Explorers in Their Own Words. Bobbs-Merrill, 1965.
The Loneliest Continent. New York Graphic Society, 1964.

Clark, Ronald W., *War Winners.* London: Sidgwick & Jackson, 1980.

Clift, A. Denis, *Our World in Antarctica.* Rand McNally, 1962.

Colbeck, Samuel C., ed., *Dynamics of Snow and Ice Masses.* Academic Press, 1980.

Cornwall, Ian, *Ice Ages: Their Nature and Effects.* Humanities Press, 1970.

Corozzi, Albert, ed., *Studies on Glaciers.* Hafner Publishing Co., 1967.

Cresswell, Robert, *The Physical Properties of Glaciers and Glaciation.* London: Hulton Ltd., 1958.

Denton, George H., and Terence J. Hughes, eds., *The Last Great Ice Sheets.* John Wiley & Sons, 1981.

Dyson, James, *The World of Ice.* Alfred A. Knopf, 1962.

Dyson, John, *The Hot Arctic.* Little, Brown, 1979.

Eggleston, Wilfrid, *Scientists at War.* Oxford University Press, 1950.

Eklund, Carl R., and Joan Beckman, *Antarctica.* Holt, Rinehart and Winston, 1963.

Embleton, Clifford, and Cuchlaine King, *Glacial Geomorphology,* Vol. 1. John Wiley & Sons, 1975.

Exploring America's Backcountry. National Geographic Society, 1979.

Fenton, Carroll and Mildred, *Giants of Geology.* Doubleday, 1952.

Field, William O., ed., *Mountain Glaciers of the Northern Hemisphere.* U.S. Government Printing Office, 1975.

Flint, R. F., *Glacial and Quaternary Geology.* John Wiley & Sons, 1971.

Fraser, Colin, *The Avalanche Enigma.* Rand McNally, 1966.

Gilluly, James, et al., *Principles of Geology*. W. H. Freeman, 1975.

Gould, Laurence McKinley, *Cold: The Record of an Antarctic Sledge Journey*. Brewer, Warren & Putnam, 1931.

Herbert, Wally, *A World of Men*. G. P. Putnam's Sons, 1969.

Hoyt, Edwin P., *The Last Explorer: The Adventures of Admiral Byrd*. John Day Co., 1968.

Huntford, Roland, *Scott and Amundsen*. G. P. Putnam's Sons, 1980.

Husseiny, A. A., ed., *Iceberg Utilization*. Pergamon Press, 1978.

Huxley, Elspeth, *Scott of the Antarctic*. Atheneum, 1978.

Imbrie, John, and Katherine Palmer, *Ice Ages: Solving the Mystery*. Enslow Publishers, 1979.

John, Brian:
The Ice Age: Past and Present. Collins, 1977.
The Winters of the World: Earth under the Ice Ages. London: David & Charles, 1979.
The World of Ice: The Natural History of the Frozen Regions. London: Orbis Publishing Ltd., 1979.

Kirwan, L. P., *A History of Polar Exploration*. W. W. Norton, 1959.

Kohnen, Heinz, *Antarktis Expedition*. West Germany: Gustav Lübbe Verlag GmbH, 1981.

Ladurie, Emmanuel Le Roy, *Times of Feast, Times of Famine: A History of Climate since the Year 1000*. Doubleday, 1971.

Langone, John, *Life at the Bottom: The People of Antarctica*. Little, Brown, 1977.

Lewis, Richard S., *A Continent for Science: The Antarctic Adventure*. Viking Press, 1965.

Lewis, Richard S., and Philip M. Smith, eds., *Frozen Future: A Prophetic Report from Antarctica*. Quadrangle Books, 1973.

Lord, Walter, *A Night to Remember*. Holt, Rinehart and Winston, 1955.

Lurie, Edward, *Louis Agassiz: A Life in Science*. University of Chicago Press, 1960.

MacDonald, Edwin A., *Polar Operations*. United States Naval Institute, 1969.

McElhinny, M. W., ed., *The Earth: Its Origin, Structure and Evolution*. Academic Press, 1979.

McElhinny, M. W., *Palaeomagnetism and Plate Tectonics*. Cambridge University Press, 1973.

McPherson, John G., *Footprints on a Frozen Continent*. Hicks Smith and Sons Ltd., 1975.

McWhinney, Mary A., ed., *Polar Research: To the Present, and the Future*. American Association for the Advancement of Science, 1978.

Matsch, Charles L., *North America and the Great Ice Age*. McGraw-Hill, 1976.

Moore, Ruth, *The Earth We Live On*. Alfred A. Knopf, 1956.

Moore, Wilfred G., *Glaciers*. London: Hutchinson, 1972.

Mountfield, David, *A History of Polar Exploration*. Hamlyn Publishing Group Ltd., 1974.

Muir, John:
The Mountains of California. Anchor Books (Doubleday), 1961.
Our National Parks. Houghton Mifflin, 1916.
Studies in the Sierra. Sierra Club, 1960.
Travels in Alaska. Houghton Mifflin, 1915.

Neider, Charles:
Beyond Cape Horn: Travels in the Antarctic. Sierra Club Books, 1980.
Edge of the World: Ross Island, Antarctica. Doubleday, 1974.

Owen, Russell, *The Antarctic Ocean*. London: Museum Press Ltd., 1948.

Paterson, W. S. B., *The Physics of Glaciers*. Pergamon Press, 1981.

Perry, Richard:

Polar Regions Atlas. Central Intelligence Agency, May 1978.

The Polar Worlds. Taplinger Publishing Co., 1973.

Ponting, Herbert G., *The Great White South*. London: Duckworth, 1922.

Ponting, Herbert, and Frank Hurley, *1910-1916 Antarctic Photographs*. St. Martin's Press, 1979.

Porter, Eliot, *Antarctica*. E. P. Dutton, 1978.

Post, Austin, and Edward R. LaChapelle, *Glacier Ice*. Toronto: University of Toronto Press, 1971.

Press, Frank, and Raymond Siever, *Earth*. W. H. Freeman, 1978.

Price, Larry W., *Mountains and Man: A Study of Process and Environment*. University of California Press, 1981.

Priestley, Sir Raymond, et al., *Antarctic Research*. Butterworths, 1964.

Quam, Louis O., ed., *Research in the Antarctic*. American Association for the Advancement of Science, 1971.

Quatermain, L. B., *South to the Pole: The Early History of the Ross Sea Sector, Antarctica*. Oxford University Press, 1967.

Raffaella, Maria, and Fiory Ceccopieri, eds., *Dal Caucaso al Himalaya 1889-1909; Vittorio Sella Fotografo Alpinista Esploratore*. Milan: Touring Club Italiano, 1981.

Robinson, Bart, *Columbia Icefield: A Solitude of Ice*. Banff: Altitude Publishing Ltd., 1981.

Rodahl, Kaare, *North: The Nature and Drama of the Polar World*. Harper & Brothers, 1953.

Rose, Lisle A., *Assault on Eternity: Richard E. Byrd and the Exploration of Antarctica 1946-47*. Naval Institute Press, 1980.

Savours, Ann, ed., *Scott's Last Voyage: Through the Antarctic Camera of Herbert Ponting*. Praeger, 1974.

Scherman, Katharine, *Daughter of Fire: A Portrait of Iceland*. Little, Brown, 1976.

Schulthess, Emil, *Antarctica*. Simon and Schuster, 1960.

Schultz, Gwen:
Ice Age Lost. Anchor Press, 1973.
Icebergs and Their Voyages. William Morrow and Co., 1975.

Scott, Robert F., *The Voyage of the 'Discovery,'* Vols. 1 and 2. Smith, Elder, & Co., 1905.

Shackleton, E. H.:
The Heart of the Antarctic, Vols. 1 and 2. J. B. Lippincott, 1909.
Nansen: The Explorer. London: H. G. & G. Witherby Ltd., 1959.

Siple, Paul, *90° South: The Story of the American South Pole Conquest*. G. P. Putnam's Sons, 1959.

Sugden, David E., and Brian S. John, *Glaciers and Landscape: A Geomorphological Approach*. Edward Arnold, 1976.

Sullivan, Walter, *Quest for a Continent*. McGraw-Hill, 1957.

Switzerland and Her Glaciers. Zurich: Kümmerly & Frey, 1981.

Tharp, Luise Hall, *Adventurous Alliance*. Little, Brown, 1959.

Thorarinsson, Sigurdur, *Glacier: Adventure on Vatnajökull, Europe's Largest Ice Cap*. Iceland Review Books, 1975.

Twain, Mark, *A Tramp Abroad*. Perennial Library, 1977.

Von Engeln, Oscar, *The Finger Lakes Region*. Cornell University Press, 1961.

Wegener, Alfred, *The Origin of Continents and Oceans*. Transl. by John Biram. Dover Publications, 1966.

Wilson, Edward, *Diary of the Terra Nova Expedition to the Antarctic 1910-1912*. Humanities Press, 1972.

Wilson, J. Tuzo, *IGY: The Year of the New Moons*. Alfred A. Knopf, 1961.

Wise, Terence, *Polar Exploration*. London: Almark Publishing Co. Ltd., 1973.

Periodicals

Alaska Tidelines, Vol. 4, No. 3, November 1981.

"Antarctica's Icy Assets." *Newsweek*, October 3, 1977.

"Antarctic Ice Pack Shrinks Dramatically." *The Washington Post*, October 21, 1981.

"Antarctic Research: Germany Joins In." *Nature*, November 27, 1980.

"Antarctic Sea Ice May Herald Ice Age." *Science News*, January 13, 1979.

Ashkenazy, Irvin, "The Rest of the Iceberg." *Oceans*, May/June 1977.

Barnes, James N., "Danger for the Antarctic." *Living Wilderness*, September 1980.

Bean, Tom, "Rivers of Ice, Rivers of Glass." *Outside*, October 1981.

Beatty, M. E., "Mountain Sheep Found in Lyell Glacier." *Yosemite Nature Notes*, December 1933.

Bentley, Charles R., "Ice-thickness Patterns and the Dynamics of the Ross Ice Shelf, Antarctica." *Journal of Glaciology*, 1979.

Bouton, Katherine, "A Reporter at Large: South of 60 Degrees South." *The New Yorker*, March 23, 1981.

Burt, Jesse C., "The Battle of the Bergs." *Natural History*, April 1956.

Burton, Robin:
"Antarctica: Rich around the Edges." *Sea Frontiers*, September/October 1977.
"Protecting the Southern Ocean." *Sea Frontiers*, May/June 1981.

Calkin, Parker E., "Subglacial Geomorphology Surrounding the Ice-free Valleys of Southern Victoria Land, Antarctica." *Journal of Glaciology*, 1974.

Chatterjee, Sankar, "The Paleoposition of Marie Byrd Land, West Antarctica." *Antarctic Journal of the United States*, 1980 Review.

Clough, John W., and B. Lyle Hansen, "The Ross Ice Shelf Project." *Science*, February 1979.

Cragin, Jim, "Tales the Ice Can Tell." *Mosaic*, September/October 1978.

"A Crustacean and Fossils Are Found by Scientists under Antarctic Ice Shelf." *The Washington Post*, December 21, 1977.

"Wie Deutschlands erst Antarktis-Station Entstand." *Diners Club Magazine*, June 1981.

Dietz, Robert S., and John C. Holden, "The Breakup of Pangaea." *Scientific American*, October 1970.

Doumani, George A., and William E. Long, "The Ancient Life of the Antarctic." *Scientific American*, September 1962.

"The Earth beneath the Poles." *Mosaic*, September/October 1978.

Ellis, William S., and James R. Holland, "Tracking Danger with the Ice Patrol." *National Geographic*, June 1968.

Flint, Richard Foster, "Glacier." *Natural History*, October 1947.

Franzier, Kendrick, "Is There an Iceberg in Your Future?" *Science News*, November 5, 1977.

Friedman, E. Imre, "Endolithic Microorganisms in the Antarctic Cold Desert." *Science*, February 1982.

"The Growing Riches of Antarctica." *New Scientist*, June 4, 1981.

Haag, William, "The Bering Strait Land Bridge." *Scientific American*, January 1962.

Hallam, A:
"Alfred Wegener and the Hypothesis of Conti-

nental Drift." *Scientific American*, February 1975.
"Continental Drift and the Fossil Record." *Scientific American*, November 1972.

Hornblower, Margot:
"The Last Untouched Continent." *The Washington Post*, February 2, 1981.
"Under World: At the Coldest Place on Earth, Nations Stake Their Claims in Frosty Silence." *The Washington Post*, February 1, 1981.

Huntford, Roland, "How Amundsen Won the Race." *Geographical Magazine*, December 1981.

Jankowski, E. J., and D. J. Drewry, "The Structure of West Antarctica from Geophysical Studies." *Nature*, May 7, 1981.

Kerr, Richard A., "Staggered Antarctic Ice Formation Supported." *Science*, July 24, 1981.

"Key to Future Locked in Antarctic Ice." *Science Digest*, June 1981.

Koenig, L. S., et al., "Arctic Ice Islands." *Journal of the Arctic Institute of North America*, July 1952.

Large, Arlen J., "South Pole Scientists Hope to Freeze Out Commercial Projects." *The Wall Street Journal*, January 7, 1981.

Lester, Marianne, "Searching for Icebergs." *The Times Magazine*, July 12, 1976.

Lipps, Jere H., et al., "Life Below the Ross Ice Shelf, Antarctica." *Science*, February 2, 1979.

McDowell, Bart, and John E. Fletcher, "Avalanche! 3,500 Peruvians Perish in Seven Minutes." *National Geographic*, June 1962.

Matthews, Samuel W.:
"Antarctica's Nearer Side." *National Geographic*, November 1971.
"What's Happening to Our Climate?" *National Geographic*, November 1976.

Mercer, J. H., "West Antarctic Ice Sheet and CO_2 Greenhouse Effect: A Threat of Disaster." *Nature*, January 1978.

Miller, Keith, "Under-Ice Volcanoes." *The Geographic Journal*, March 1979.

Mitchell, Barbara, "The Politics of Antarctica." *Environment*, January/February 1981.

Mitchell, Barbara, and Lee Kimball, "Conflict over the Cold Continent." *Foreign Policy*, Summer 1979.

Mitchell, J. Murray, Jr., "Carbon Dioxide and Future Climate." *Environmental Data Service*, March 1977.

Moore, Tui De Roy, "The Day of the Glacier." *Adventure Travel*, September/October 1981.

Moulton, Kendall N., and Richard L. Cameron, "Bottle-Green Iceberg Near the South Shetland Islands." *Antarctic Journal of the United States*, June 1976.

Olson, Suzanne, "Solar Tracks in the Snow." *Science News*, November 15, 1980.

Parker, Bruce C., et al., "Modern Stromatolites in Antarctic Dry Valley Lakes." *BioScience*, October 1981.

"Peru, Avalanche." *Newsweek*, January 22, 1962.

Petit, Charles, "Antarctica: Freeze-Dried Desert." *Science Digest*, May 1981.

Platt, Rutherford, "A Visit to the Living Ice Age."

National Geographic, April 1957.

"Poland Freezes Antarctic Research." *Nature*, January 8, 1981.

Porter, Eliot, "Images of Antarctica." *Living Wilderness*, June 1979.

Quigg, Philip W., "One World." *Audubon*, July 1978.

Reinhold, Robert:
"Antarctica Yields First Land Mammal Fossil." *The New York Times*, March 21, 1982.
"Antarctic Explorers Shift Goal to Hidden Resources." *The New York Times*, December 20, 1981.
"As Others Seek to Exploit Antarctic, U.S. Takes the Scientific Approach." *The New York Times*, December 21, 1981.

"Science in Antarctica: A Summary of National Activities." *Antarctic Journal of the United States*, March 1981.

Smith, W. J., "The International Ice Patrol." *Sperryscope*, 1966.

Sullivan, Walter:
"Antarctic Laboratories: The Many Uses of Cold." *The New York Times*, February 5, 1978.
"Drill Pierces Ice Shelf, Opening 'Lost World' to Scientists." *The New York Times*, December 9, 1977.
"Fresh Clues to a Green Antarctica Found under Ice." *The New York Times*, February 11, 1979.
"Ice in Antarctica Found to Wax and Wane." *The New York Times*, December 27, 1976.
"Radar Maps a Rugged Land beneath Antarctica." *The New York Times*, July 31, 1978.
"Radar to Help Trace Profile of Antarctica." *The New York Times*, November 7, 1969.
"Scientists Reviving Speculations on Climate and Slipping Antarctic Ice." *The New York Times*, March 9, 1980.

Swithinbank, Charles W., "Ice Movement of Valley Glaciers Flowing into the Ross Ice Shelf, Antarctica." *Science*, August 9, 1963.

"Tales the Ice Can Tell." *Mosaic*, September/October 1978.

Thomas, Robert H., et al., "Effect of Climatic Warming on the West Antarctic Ice Sheet." *Nature*, February 1, 1979.

Wallace, William J., "Habakkuk." *Warship*, April 1981.

Washburn, A. L., "Focus on Polar Research." *Science*, August 8, 1980.

Wasmund, Erich, "Report on the Corpse-Wax in the Mountain Sheep Found in the Ice of the Lyell Glacier." *Yosemite Nature Notes*, March 1938.

"Weather from the Ends of the Earth." *Mosaic*, September/October 1978.

Wharton, Robert A., et al., "Biogenic Calcite Structures Forming in Lake Fryxell, Antarctica." *Nature*, February 4, 1982.

Wilson, A. T., "Origin of Ice Ages: An Ice Shelf Theory for Pleistocene Glaciation." *Nature*, January 11, 1964.

Wilson, J. Tuzo, "Continental Drift." *Scientific American*, April 1963.

Wold, Bjørn and Gdnnar Østrem:
"Morphological Activity of a Diverted Glacier Stream in Western Norway." *GeoJournal*, 1979.
"Subglacial Constructions and Investigations at Bondhusbreen, Norway." *Journal of Glaciology*, Vol. 23, No. 89.

Young, Patrick, "Unique Life Forms at the End of the Earth." *Smithsonian*, November 1981.

Zegarelli, Philip E., "Antarctica." *Focus*, September/October 1978.

Other Publications

Annals of Glaciology. Proceedings of the Conference on Use of Icebergs: Scientific and Practical Feasibility, Cambridge, U.K., April 1-3, 1980. International Glaciological Society, Vol. 1, 1980.

"The Antarctic and Its Geology." U.S. Department of the Interior Geological Survey pamphlet, 1978.

Bentley, C. R., "Variations in Valley Glacier Activity in the Transantarctic Mountains as Indicated by Associated Flow Bands in the Ross Ice Shelf." Proceedings of the Canberra Symposium, December 1979, IAHS Publication No. 131.

Holdsworth, Gerald, "An Examination and Analysis of the Formation of Transverse Crevasses, Kaskawulsh Glacier, Yukon Territory, Canada." Unpublished thesis, The Ohio State University, 1965.

"International Ice Patrol." United States Coast Guard pamphlet, 1977.

Kazarian, Ralph:
"Scientists to Explore Antarctic Islands to Study Riddle of Continental Drift." United States Antarctic Program press release, September 4, 1981.
"Scientists to Study Krill in Antarctic Waters, Establish Solar Telescopes at South Pole." United States Antarctic Program press release, September 22, 1980.

Mitchell, Barbara, and Jon Tinker, "Antarctica and Its Resources." EARTHSCAN Press Briefing Document No. 21, September 1979.

Morales, Benjamin, "The Huascarán Avalanche in the Santa Valley, Peru." *Reports and Discussions of the International Symposium on Scientific Aspects of Snow and Ice Avalanches, April 5-10, 1965, Davos, Switzerland*. International Association of Scientific Hydrology, Publication No. 69, 1966.

Potter, Neal, "Natural Resource Potentials of the Antarctic." The American Geographical Society, Occasional Publication No. 4, 1969.

Sater, John E., ed., *Arctic Drifting Stations: A Report on Activities Supported by the Office of Naval Research*. Proceedings of the Symposium held at Airlie Conference Center, Warrenton, Virginia, April 12-15, 1966, under the auspices of The Arctic Institute of North America and the Office of Naval Research, U.S. Navy, November 1968.

Sharp, Robert P., *Glaciers*. Condon Lectures, Oregon State System of Higher Education, Eugene, Oregon, 1960.

U.S. Bureau of Mines, *Minerals Yearbook, 1980*. Washington Government Printing Office, 1980.

ACKNOWLEDGMENTS

For their help in the preparation of this book the editors wish to thank: **In Argentina:** Buenos Aires—Juan Gomez. **In Austria:** Innsbruck—Dr. Michael Kuhn, University of Innsbruck. **In Canada:** St. Johns, Newfoundland—Angus Bruneau, Bruneau Resources Management, Ltd., NORDCO; Ottawa—C. Simon Ommanney, Director, Snow and Ice Division, National Hydrology Research Institute; Vancouver—G. K. C. Clarke, University of British Columbia. **In Denmark:** Charlottenlund—Hans Ebbesen, Arktisk Institute; Copenhagen—Dr. Willi Dansgaard, Geophysical Isotope Laboratory, University of Copenhagen; Tony Higgins and Anker Weidick, The Geological Survey of Greenland. **In France:** Mantes-La-Ville—Jean-Paul Peulvast; Neuilly-sur-Seine—Geneviève Viollet-Leduc; Paris—Claude Bellarbre and Marjolaine Matikhine, Musée de la Marine; Pierre Gérin. **In Great Britain:** Buckinghamshire—R. W. Shirley; Cambridge—C. Holland and H. G. R. King, Scott Polar Research Institute; C. Swithinbank, British Antarctic Survey; London—H. G. Bilcliffe, F. Herbert, Nada Tosic and Shane Wesley-Smith, Royal Geographical Society; R. T. Williams, Department of Prints and Drawings, British Museum; Marjorie Willis, BBC Hulton Picture Library; Sheffield—Professor K. J. Miller, University of Sheffield; Staffordshire/E. Derbyshire—University of Keele. In Italy: Biella—Antonio Canevarolo and Lodovico Sella, Istituto di Fotografia Alpina. In Japan: Nagoya—Dr. Keiji Higuchi, Water Research Institute, Nagoya University. In the Netherlands: Amsterdam—The Rijksmuseum. In Northern Ireland: Belfast—W. B. Whalley, Queen's University of Belfast. In Norway: Oslo—Jon Ove Hagen, Department of Geography, University of Oslo; Norwegian Polar Research Institute; Bjørn Wold, Hydrological Division, Norwegian Water Resources and Electricity Board. In Peru: Lima—Cesar Morales Arnao, National Recreation Institute; Servicio Aerofotografico Nacional. In South Africa: Johannesburg — Nigel Brown; Peter Johnson. In Switzerland: Neuchâtel—Eric Jeannet, Université de Neuchâtel; Zurich—Eleanora Frizzoni, S.N.T.O. In the United States: Alaska—(Anchorage) John Schindler; (Fairbanks) Dr. Larry Mayo; California—(Santa Monica) John Hult; (Woodland Hills) Neil Ostrander; (Yosemite National Park) Mary Volcecka, National Park Service Research Library; District of Columbia—Dr. Richard Cameron, Winifred M. Reuning and Frank Williamson, Division of Polar Programs, National Science Foundation; Richard Muldoon and Walt Seelig, Public Relations Office, National Science Foundation; Margot Hornblower, *The Washington Post;* John Sader; Alison Wilson, Archives Technician, Center for Polar and Scientific Archives, National Archives; Illinois—(Urbana) Art Devries, University of Illinois; Maine—(Orono) Dr. Terence Hughes, Institute of Quaternary Studies, University of Maine; Maryland—(Bethesda) Albert Crary; Massachusetts—(Boston) Dr. Bradford Washburn, Chairman, Museum of Science; (Cambridge) Dr. James Fullmer, Department of Meteorology, Massachusetts Institute of Technology; John Waterhouse, Massachusetts Institute of Technology Museum; (Great Barrington) Dr. William O. Field Jr.; (Indian Orchard) Edward Kamuda; Minnesota—(Minneapolis) Professor Roger Hooke, Department of Geology and Geophysics, University of Minnesota; Nebraska—(Lincoln) Bruce Koci and Karl C. Kuivenen, Director, Polar Ice Coring Office, University of Nebraska; New Hampshire—(Durham) Dr. Paul Mayewski, Department of Earth Sciences, University of New Hampshire; (Hanover) Malcolm Mellor and Dr. Wilford Weeks, U.S. Army Cold Regions Research and Engineering Laboratory; New York—(Amherst) Parker Calkin and Dr. Michael Herron, Department of Geological Sciences, State University of New York at Buffalo; (New York) Lieutenant (j.g.) Michael Cicalese and Commander Joseph J. McClelland Jr., International Ice Patrol, U.S. Coast Guard; (Oneonta) Dr. Jay Fleisher, Department of Geology, State University of New York at Oneonta; Ohio—(Columbus) Dr. David Elliot, Dr. John Mercer, Dr. Lonnie Thompson and Dr. Ian Whilliams, Institute of Polar Studies, The Ohio State University; Virginia—(Blacksburg) Gordon Love, Department of Geological Sciences, Virginia Polytechnical and State University; (Burke) Commander Arthur Shepard; (Reston) Dr. Richard S. Williams Jr., U.S. Geological Survey; Washington—(Seattle) Kevin Schafer; (Tacoma) David R. Hirst, Dr. Mark Meier and Austin Post, Glaciology Division, U.S. Geological Survey; Wisconsin—(Madison) Dr. Charles Bentley, Geophysical and Polar Research Center, and Gwen Schultz, Department of Geography, University of Wisconsin. In West Germany: Münster—Professor Dr. H. Kohnen.

The editors also wish to thank Jeanne Abbott, Anchorage; Bob Gilmore, Auckland; Enid Farmer, Boston; Nina Lindley, Buenos Aires; Lois Lorimer, Copenhagen; Robert Kroon, Geneva; Lance Keyworth, Helsinki; Peter Hawthorne, Johannesburg; Tom Loayza, Lima; John Dunn, Melbourne; Felix Rosenthal, Moscow; Warren Vieth, Oklahoma City; Cal Abraham, Santiago; Mary Johnson, Stockholm; Ed Reingold, Tokyo.

The index was prepared by Gisela S. Knight.

INDEX